Developing Algebraic Thinking

by Don Balka

Didax Educational Resources®

Copyright © 2005 by Didax, Inc., Rowley, MA 01969. All rights reserved.

Limited reproduction permission: The publisher grants permission to individual teachers who have purchased this book to reproduce the blackline masters in quantities sufficient for that teacher's classroom.

Reproduction of any part of this book for use by other than the puchaser–an entired school or school district–is strictly prohibited.

Except as noted above, no part of this publication may be reproduced, stored in a retrieval system, or transmitted in any form or by any means–electronic, mechnical, recording, or otherwise–without the prior written consent of the publisher.

Printed in the United States of America.

Order Number 2-5251
ISBN 1-58324-216-3

A B C D E F 09 08 07 06 05

395 Main Street
Rowley, MA 01969
www.worldteacherspress.com

DEVELOPING ALGEBRAIC THINKING

Foreword

Over the past few years, many mathematics educators have directed their interests towards the development of algebraic thinking in students. As more and more school districts move in the direction of algebra for all, teachers need to be able to provide their students with the knowledge and skills necessary for success in a mathematics class where the focus on algebra is considerably different than the traditional course of study.

What is algebraic thinking? Although there does not appear to be one agreed-upon definition, several common threads appear in the related literature.

- *Exploring and conjecturing about patterns*
- *Formalizing patterns*
- *Verbalizing relationships*
- *Making generalizations*
- *Symbolizing relationships*
- *Working with functions*
- *Making connections between real world situations and algebraic statements*

Together, all of these interrelated components form a framework for algebraic thinking. This type of thinking "embodies the construction and representation of patterns and regularities, deliberate generalization, and most important, active exploration and conjecture (Chambers, 1994, 85)." Problem solving and problem solving strategies, as described in the *NCTM Principles and Standards for School Mathematics (2000)* permeate the activities.

Contents

Teacher's Notes	3–8
Place Value Picks	9–22
Same Sums	23–34
Primes and Composites	35–44
Divisibility Rules	45–60
Locker Numbers	61–68
Exciting Components	69–78
Plus and Times	79–86
Shapes	87–98
Pentomino Puzzles	99–108
Alphabet Algebra	109–142
Numbered Tiles	143

PROBLEM SOLVING AND NUMBER TILES

Teacher's Notes

Developing Algebraic Thinking provides activities to meet some of the common threads cited in the *Foreword*. Using number tiles 0 through 9, students are given opportunities to solve a variety of problems involving algebra ideas. The activities lend themselves to finding solutions by individuals or small groups, followed by class discussion of the observable patterns or relationships and the related algebra. With teacher direction, the emerging patterns in many of the activities provide an ideal opportunity for discussion of number sense, place value, basic number theory, variables, and simple algebraic expressions and equations.

Most of the activities in the book have more than one correct solution. Sometimes, number tiles may be rearranged to give the same result. Here are two different examples illustrating this point.

Example 1: Use any 6 tiles to create two 3-digit numbers whose difference is the least.

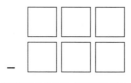

There are five solutions to this problem, all giving a difference of 3.

3 0 1	4 0 1	5 0 1	6 0 1	7 0 1
− 2 9 8	− 3 9 8	− 4 9 8	− 5 9 8	− 6 9 8
3	3	3	3	3

Example 2: Use any 6 tiles to create two 2-digit addends and a 2-digit sum.

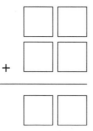

With a sum of 98, four different solutions exist.

4 2	4 6	4 1	4 7
+ 5 6	+ 5 2	+ 5 7	+ 5 1
9 8	9 8	9 8	9 8

DEVELOPING ALGEBRAIC THINKING

PROBLEM SOLVING AND NUMBER TILES

Teacher's Notes

In many cases, as the first example suggests, students are asked to find the greatest or least possible answer. These situations provide an ideal time to initiate class discussion of problem solving strategies for finding a particular solution. As students work through the number tile activities, here are some of the strategies that might be applied:

- *Guess and check*
- *Look for a pattern*
- *Write an equation*
- *Solve a simpler problem*
- *Work backwards*
- *Estimating*

Below are sample activities to use as warm-ups for your class.

Activity 1 Use any 6 tiles to create a 2-digit factor and a 1-digit factor with a correct 3-digit product.

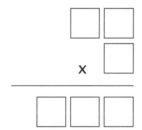

The directions make the problem open-ended; any correct solution is acceptable.

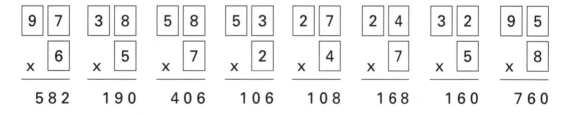

The least possible product: should have a 1 in the hundreds place. Therefore, neither factor can contain a 1. Doubling a number in the fifties provides a good start: 53 x 2 = 106.

The greatest product: requires some simple number sense work to begin. Since 80 x 9 = 90 x 8 = 720 and 9 and 8 are the greatest digits for the tens place in the one factor, it appears that the hundreds digit in the product will be 7. After some testing, 95 x 8 = 760 gives the greatest product.

PROBLEM SOLVING AND NUMBER TILES

Teacher's Notes

Activity 2 Use any 6 number tiles 0 through 9 to fill in the squares so that the sum of each lines is the same.

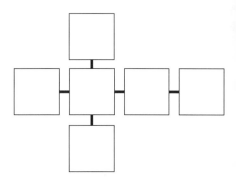

Let students find solutions for the problem. Post them on the chalkboard or the overhead projector, and discuss patterns that emerge.

```
   8              4              8
6 9 4 5        2 3 0 8        9 5 2 1
   7              6              4
Sum = 24       Sum = 13       Sum = 17
```

Now assign variables to each square and look at the equations that are created.

$$e$$
$$a\ b\ c\ d$$
$$f$$

$a + b + c + d = e + b + f$ or $a + c + d = e + f$

The algebra shows that b can be any digit; each time one solution is found, four others can be found by replacing the tile represented by b as shown below.

```
   2          2          2          2          2
1 0 3 4    1 5 3 4    1 7 3 4    1 8 3 4    1 9 3 4
   6          6          6          6          6
```

The least sum that satisfies the equation $a + c + d = e + f$, is 5: $0 + 1 + 4 = 2 + 3$. Therefore, the least sum for the problem is 10:

```
   2
0 5 1 4
   3
```

The greatest sum for the problem is 22.

Activity 3 Use all 10 number tiles 0 through 9 to create a 10-digit number divisible by 3.

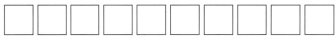

This is a simple problem that focuses on the divisibility test for 3. Since the sum of the digits 1 through 9 is 45, and 45 is divisible by 3, any arrangement of the digits other than 0 in the first place provides a solution.

PROBLEM SOLVING AND NUMBER TILES — Teacher's Notes

The book contains 10 sections, each with a part entitled "Looking at the Algebra." In the first section, "Place Value Picks," students use number sense ideas involving place value to create various sums, differences, and products. "Same Sums" requires students to separate the number tiles into a designated number of groups so that each group has the same sum. As the activity progresses, teachers can introduce Gauss' formula for the sum of the first n consecutive counting numbers. Creating prime numbers and composite numbers of various lengths is the focus in "Primes and Composites." Number sense ideas continue in the next section, "Divisibility Rules," as students create numbers satisfying divisibility criteria for 2, 3, 4, 5, 6, 8, 9, and 10. "Locker Numbers" provides the first activities where variables can be incorporated into classroom discussion. Still, number sense ideas—even/odd numbers, multiples, factors—play an important role in finding solutions.

Activities in "Exciting Exponents" stress basic rules of exponents, along with number theory ideas. "Plus and Times" provides students opportunities to look for simple arithmetic patterns with sums or products, and then look at the associated algebra. Various polygons are made into number tile problems in the "Shapes" section. From triangles to pentagons, and finally to a box, the number of variables and the number of equations increases. "Pentomino Puzzles" uses various pentomino shapes as number tile puzzles, creating lines of digits whose sums will be the same. The last section, "Alphabet Algebra," contains number tile puzzles for the entire alphabet. As in other sections, each line (or curve, in some cases) has the same sum. The algebra is more difficult; however, interesting relationships among the digits do appear.

Let your students create their own number tile problems, expressing solutions using variables if possible, discussing patterns, verbalizing relationships, and making conjectures. What hexominoes can be made into number tile problems? What interesting addition, subtraction, multiplication, or division problems can be made? Developing algebraic thinking, using only number tiles, can be an exciting, challenging, and rewarding experience for your students.

Reference: Chambers, Donald L. "The Right Algebra for All." Educational Leadership 51 (March 1994): 85 – 86.

NCTM, *Principles and Standards for School Mathematics,* Reston, VA: NCTM, 2000.

TEACHER'S NOTES

Place Value Picks

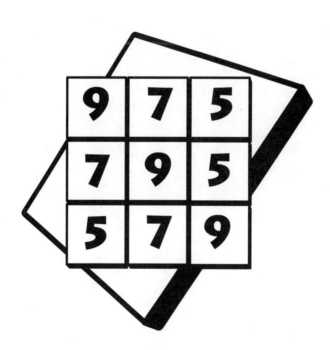

Place Value Picks

Teacher's Notes

Introduction All of the problems in this section focus on place value concepts. *Place Value Picks* involves finding solutions to a variety of multiplace addition and subtraction problems using number tiles. Students are directed to find the greatest or least possible sum or difference for each problem. Most often, number tiles are not used in the answers; however, problems are included where the tiles are also a part of the answer.

As students begin to solve these problems, have them first find any solution to start, regardless of whether the solution is the greatest or least. This acceptance by the teacher of their success will provide some confidence as they seek the greatest or least result.

Problems involving number tiles in the answer are more difficult and may require additional time. Although basic ideas of place value still hold, trading (carrying) creates unusual situations.

Looking at the Algebra Number sense, as a part of algebraic thinking, becomes the key to solving the problems in this section. For a greatest sum, greatest digits are placed in the places to the left. For a least sum, least digits are placed in the places to the left. However students must be careful with the digit 0.

Place Value Picks 1 The first problem on *Place Value Picks 1* is straightforward. To obtain the greatest sum, the greatest digits need to be placed in the squares for the highest place value. Although rearrangement of the number tiles is possible, the greatest sum is still 1839.

In finding the least sum, 0 cannot be used in the hundreds place; therefore 1 and 2 must be used. The 0 tile is then used in the tens place. Again, rearrangement of number tiles is possible. The least sum remains 339.

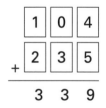

Place Value Picks 2 The strategies for finding the greatest and least sums for *Place Value Picks 2* are the same as those for *Place Value Picks 1*.

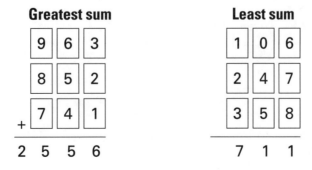

DEVELOPING ALGEBRAIC THINKING

Place Value Picks

Teacher's Notes

Place Value Picks 3

For *Place Value Picks 3*, the strategies are the same.

Greatest sum	Least sum	Greatest sum	Least sum
9 7 5 3	1 0 4 6	9 7 5 3 1	1 0 4 6 8
+ 8 6 4 2	+ 3 3 9 8	+ 8 6 4 2 0	+ 2 3 5 7 9
1 8 3 9 5	3 4 0 3	1 8 3 9 5 1	3 4 0 4 7

As mentioned earlier, when number tiles are used in the answers finding solutions becomes more difficult. Rather than focusing on the addends, one must often focus on the sum. Therefore, working backwards becomes an acceptable problem solving strategy.

Place Value Picks 4

In *Place Value Picks 4*, a desirable greatest sum would have a 9-tile in the tens place, followed by an 8–tile in the ones place. With these digits to start and using a "Guess and check" strategy, several solutions can be found.

Greatest sum

```
    7 5        6 1
+   2 3    +   3 7
   ----       ----
    9 8        9 8
```

The least sum is much easier to find. The 3-tile must go in the tens place. This can only be accomplished using the 1-tile and 2-tile in the tens place of the two addends. With the 4-tile and 5-tile in ones place of the addends, the 9-tile would be needed in the ones place of the sum. The solution becomes:

Least sum

```
    1 4
+   2 5
   ----
    3 9
```

DEVELOPING ALGEBRAIC THINKING

Place Value Picks Teacher's Notes

Place Value Picks 5 For the greatest sum in *Place Value Picks 5*, we know that the digit in the hundreds place of the sum is 1. Likely choices in the tens place are 9 and 8. Using "Guess and check," a solution can be found:

Greatest sum

```
   8 4
 + 9 2
 -----
 1 7 6
```

The least sum also has a 1 in the hundreds place. A desirable least sum would have the 0-tile in the tens place and the 2-tile in the ones place, thus requiring the sum of the digits to be 10 with trading.

Least sum

```
   6 7
 + 3 5
 -----
 1 0 2
```

Place Value Picks 6 In *Place Value Picks 6*, a desirable greatest sum would have a 9-tile in the hundreds place, followed by an 8-tile in the tens place. With these digits to start and again using a "Guess and check" strategy, a solution can be found.

Greatest sum

```
   7 4 5
 + 2 3 6
 -------
   9 8 1
```

Finding the least sum in number tiles follows a similar pattern as before. The 3-tile must go in the hundreds place. This can only be accomplished using the 1-tile and 2-tile in the hundreds place of the two addends. With the 0-tile and 4-tile in the tens place of the addends, the 5-tile would be needed in the tens place of the sum. This requires the 7-tile and 9-tile in the ones place of the addends and the 6-tile in the ones place of the sum. The solution becomes:

Least sum

```
   1 0 7
 + 2 4 9
 -------
   3 5 6
```

DEVELOPING ALGEBRAIC THINKING

Place Value Picks Teacher's Notes

Place Value Picks 7

For the greatest sum in *Place Value Picks 7*, we know that the digit in the thousands place of the sum is 1. Likely choices of 9 and 8 in the hundreds place again cause duplicate digits in other places. Using "Guess and check," a solution can be found:

Greatest sum

```
  8 4 9
+ 7 5 3
-------
1 6 0 2
```

The least sum also has a 1 in the thousands place. A desirable least sum would have the 0-tile in the hundreds place and a 2-tile in the tens place, thus requiring the sum of the digits in the hundreds place to be 10 with trading.

Least sum

```
  4 8 9
+ 5 3 7
-------
1 0 2 6
```

Place Value Picks 8

Subtraction is involved in *Place Value Picks 8*. To find the greatest difference, merely place the greatest digits 9, 8, and 7 in the known sum (minuend) in decreasing order. The least digits 0, 1, and 2 need to be reordered in the known addend (subtrahend).

Greatest difference

```
  9 8 7
- 1 0 2
-------
  8 8 5
```

The least difference can be found in five different ways, although only the hundreds digits change.

Least difference

```
  7 0 1     6 0 1     5 0 1     4 0 1     3 0 1
- 6 9 8   - 5 9 8   - 4 9 8   - 3 9 8   - 2 9 8
-------   -------   -------   -------   -------
    3         3         3         3         3
```

Place Value Picks 1

Using 6 number tiles, create two 3-digit numbers so that the sum of the numbers is the greatest.

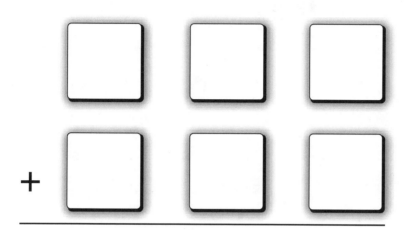

Now, using 6 number tiles, create two 3-digit numbers so that the sum of the numbers is the least.

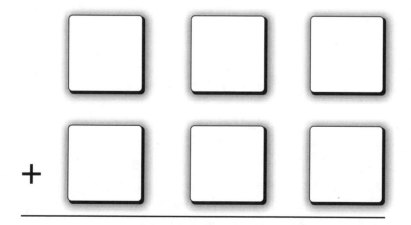

Place Value Picks 2

Using 9 number tiles, create three 3-digit numbers so that the sum is the greatest.

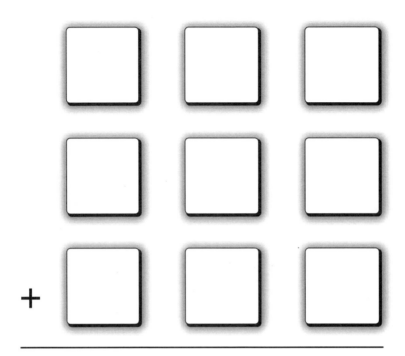

Now, using 9 number tiles, create three 3-digit numbers in the squares above so that the sum is the least.

Record your numbers and sums in the space provided below.

Greatest sum

Least sum

Place Value Picks 3

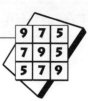

Using 8 number tiles, create two 4-digit numbers so that the sum is the greatest. Create two 4-digit numbers so that the sum is the least.

Using 10 numbers tiles, create two 5-digit numbers so that the sum is the greatest. Create two 5-digit numbers so that the sum is the least.

DEVELOPING ALGEBRAIC THINKING

Place Value Picks 4

Using 6 number tiles, create two 2-digit numbers so that the sum is also in number tiles.

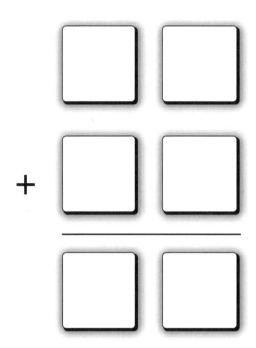

What is the greatest sum that you can find? What is the least sum that you can find?

Record your results in the spaces provided.

Greatest sum

Least sum

Place Value Picks 5

Using 7 number tiles, create two 2-digit numbers so that the sum is also in number tiles.

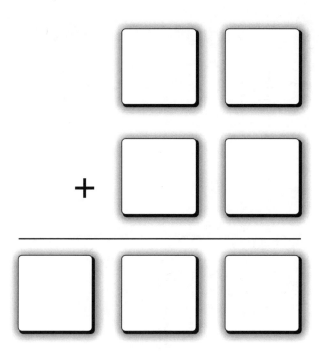

What is the greatest sum that you can find? What is the least sum that you can find?

Record your results in the spaces provided.

Greatest sum   **Least sum**

Place Value Picks 6

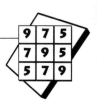

Using 9 number tiles, create two 3-digit numbers so that the sum is also in number tiles.

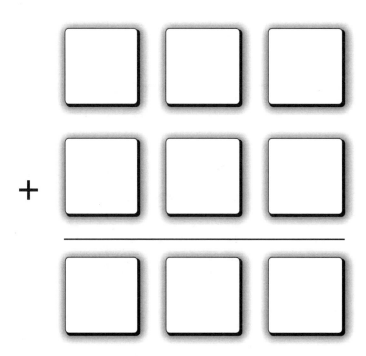

What is the greatest sum that you can find? What is the least sum that you can find?

Record your results in the spaces provided.

Greatest sum

Least sum

Place Value Picks 7

Using all 10 number tiles, create two 3-digit numbers so that the sum is also in number tiles.

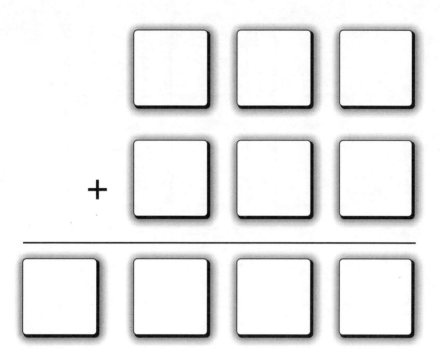

What is the greatest sum that you can find? What is the least sum that you can find?

Record your results in the spaces provided.

Greatest sum   **Least sum**

Place Value Picks 8

Using 6 number tiles, create two 3-digit numbers so that the difference of the numbers is the greatest.

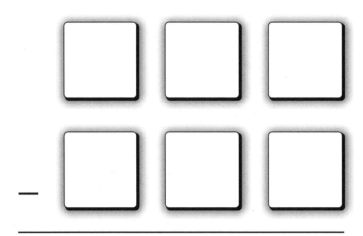

Now, using 6 number tiles, create two 3-digit numbers so that the difference of the numbers is the least.

Same Sums

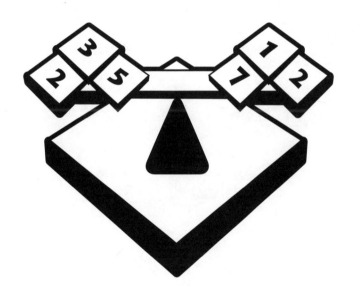

Same Sums   Teacher's Notes

Introduction

Same Sums provides an opportunity for teachers to introduce or focus on the mathematician Gauss' formula for the sum of the first "n" consecutive positive integers.

In a historical note, a story is told that young Gauss was given the following problem to solve: find the sum of the first one hundred counting numbers; that is,

1 + 2 + 3 + 4 + 5 ... + 98 + 99 + 100

His solution was simple:

$$\begin{array}{c} 1 + 2 + 3 + 4 + 5 + \ldots + 98 + 99 + 100 \\ 100 + 99 + 98 + 97 + 96 = \ldots + 3 + 2 + 1 \\ \hline 101 + 101 + 101 + 101 + 101 \quad \ldots \quad + 101 + 101 + 101 \end{array}$$

There are one hundred 101s, but this is twice the sum. So, the sum S is half of this.

$$S = \frac{100\,(101)}{2}$$

Looking at the Algebra

Looking at Gauss' technique above, we discover that the basic formula for the sum of the first n counting numbers is

$$S = \frac{n(n+1)}{2}$$

where n is the last term or number in the series.

The sums which are derived from the formula are, in fact, the triangular numbers:

1, 3, 6, 10, 15, 21, 28, 36, 45, 55, ...

i.e.   1 = 1
        2 = 1 + 2
        6 = 1 + 2 + 3 ... and so on.

These numbers appear in many places throughout mathematics.

For each page in this section, students are required to separate the number tiles into a given number of groups so that the sum of the digits in each group is the same. The number of tiles in each group is not necessarily the same; obviously the 0-tile can be placed in any group.

Same Sums

Teacher's Notes

The following table shows results using various groupings of tiles.

Tiles used	Sum of all digits	Number of groups	Sum of each group	Possible solution
0 – 3	6	2	3	0 3; 1 2
0 – 4	10	2	5	0 4 1; 2 3
0 – 5	15	3	5	0 5; 1 4; 2 3
0 – 6	21	3	7	0 1 6; 2 5; 4 3
0 – 7	28	2	14	0 2 3 4 5; 1 6 7
0 – 7	28	4	7	0 7; 1 6; 2 5; 4 3
0 – 8	36	2	18	0 1 2 4 5 6; 3 7 8
0 – 8	36	3	12	0 4 8; 1 5 6; 2 3 7
0 – 8	36	4	9	0 1 8; 2 7; 3 6; 4 5
0 – 9	45	3	15	0 1 2 3 4 5; 6 9; 7 8
0 – 9	45	5	9	0 9; 1 8; 2 7; 3 6; 4 5

Same Sums can be extended by using additional number tiles with 1 11, 12 … In using just 0 though 9 or extending the problems, a variety questions can be posed.

- What is the sum of all the numbers that you are using?

   This provides an opportunity to look at Gauss' formula.

- How many groups could we make that have the same sum?
   The number of groups is a divisor of the sum; however not all diviso provide a solution.

- What is the sum for each group?

   In using 0 through 10, the sum is 55. We could only have 5 group each with a sum of 11: 0 1 10; 2 9; 3 8; 4 7; 5 6

- When can we have consecutive numbers for the sums of ea group?

   If the number of groups is odd, then we can separate the number tiles into groups so that the sums will be consecutive.

- When can we have consecutive even numbers for the sums of ea group?

- When can we have consecutive odd numbers for the sums of ea group?

Same Sums

Teacher's Notes

If the number of groups is even and the equal sum is odd, then the number tiles can be separated into groups so that the sums will be consecutive even numbers.

If the number of groups is even and the equal sum is even, then the number tiles can be separated into groups so that the sums will be consecutive odd numbers.

CONSECUTIVE?

Tiles used	Sum of all digits	Number of groups	Sum of each group	Consecutive?
0 – 3	6	2	3	Even: 2 4
0 – 4	10	2	5	Even: 4 6
0 – 5	15	3	5	4 5 6
0 – 6	21	3	7	6 7 8
0 – 7	28	2	14	Odd: 13 15
0 – 7	28	4	7	4 6 8 10
0 – 8	36	2	18	Odd: 17 19
0 – 8	36	4	9	Even: 6 8 10 12
0 – 9	45	3	15	14 15 16
0 – 9	45	5	9	7 8 9 10 11

Same Sums

Same Sums 1   Use the number tiles 0 through 3. Separate them into 2 groups so that t sum of the digits in each group is the same. The number of tiles in ea group does not have to be the same.

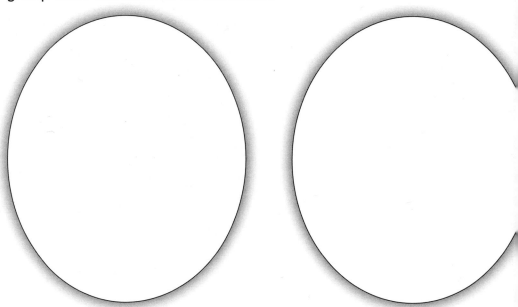

Same Sums 2   Use the number tiles 0 through 4. Separate them into 2 groups so that t sum of the digits in each group is the same. The number of tiles in ea group does not have to be the same.

Same Sums

Same Sums 3   Use the number tiles 0 through 5. Separate them into 3 groups so that the sum of the digits in each group is the same. The number of tiles in each group does not have to be the same.

Same Sums 4   Use the number tiles 0 through 6. Separate them into 3 groups so that the sum of the digits in each group is the same. The number of tiles in each group does not have to be the same.

Same Sums

Same Sums 5 Use the number tiles 0 through 7. Separate them into 2 groups so that the sum of the digits in each group is the same. The number of tiles in each group does not have to be the same.

Same Sums 6 Use the number tiles 0 through 8. Separate them into 2 groups so that the sum of the digits in each group is the same. The number of tiles in each group does not have to be the same.

Same Sums

Same Sums 7 Use the number tiles 0 through 7. Separate them into 4 groups so that the sum of the digits in each group is the same. The number of tiles in each group does not have to be the same.

Same Sums 8 Use the number tiles 0 through 8. Separate them into 4 groups so that the sum of the digits in each group is the same. The number of tiles in each group does not have to be the same.

Same Sums

Same Sums 9 Use the number tiles 0 through 9. Separate them into 3 groups so that the sum of the digits in each group is the same. The number of tiles in each group does not have to be the same.

Same Sums

Same Sums 10 Use the number tiles 0 through 9. Separate them into 5 groups so that the sum of the digits in each group is the same. The number of tiles in each group does not have to be the same.

Same Sums

Consecutive Sums 11 Use the number tiles 0 through 9. Separate them into 3 groups so that the 3 sums are consecutive numbers. The number of tiles in each group does not have to be the same.

Primes and Composites

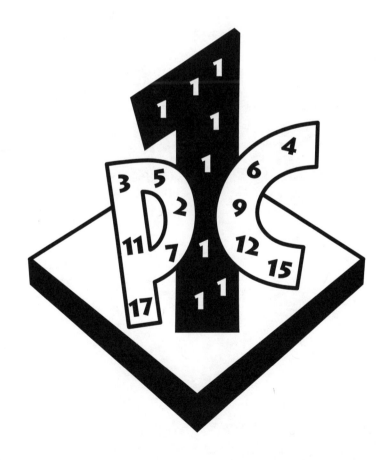

PRIMES AND COMPOSITES

Teacher's Notes

Introduction — The number tile activities in this section are designed to provide emphasis on primes and composites by having students create a variety of each—one, two, three, or four digits in length. Number theory ideas and problem solving strategies are an integral part of the activities.

Looking at the Algebra — Work with prime numbers and composite numbers serves as a basis for later work with factoring. As students progress through the activities, several key points need to be brought to their attention.

- *One is neither prime nor composite.*
- *The only even prime number is the one-digit number 2; all other prime numbers are odd.*
- *Prime numbers, other than 2, must have an odd digit in the ones place.*
- *The only prime number with the digit 5 in the ones place is the number 5; all other numbers with a 5 in the ones place are multiples of 5.*
- *Multiplace prime numbers, therefore, have 1, 3, 7, or 9 in the ones place.*
- *Just because the ones digit is odd, does not guarantee that the number is prime. Divisibility tests play an important role in determining the primeness of the number.*
- *A number is even when the ones digit is 0, 2, 4, 6, or 8 and therefore not prime (other than 2).*

Lots of Primes, Lots of Composites — In *Lots of Primes, Lots of Composites*, eight number tiles are used to create four 2-digit prime numbers, followed by creation of four composite 2-digit numbers. Here are some possible solutions:

Lots of Primes		Lots of Composites	
29	83	90	18
41	61	63	39
53	59	75	65
67	47	92	74

Primes and Composites — In *Primes and Composites*, students are first required to use all 10 number tiles to create three 2-digit prime numbers and two 2-digit composite numbers. Secondly, they are required to create two 2-digit prime numbers and three 2-digit composite numbers. Both tasks have numerous solutions and focus on the points listed above. Here are some possible solutions:

23	60	90	71
47	80	65	23
59		84	

© Didax Educational Resources · **DEVELOPING ALGEBRAIC THINKING**

PRIMES AND COMPOSITES

Teacher's Notes

More Primes *More Primes* involves using all 10 number tiles to create four primes—two 3-digit primes and two 2-digit primes. Other than using a calculator, a "primeness check" is a valuable tool on this task. Simply stated, the check says: If any prime number less than or equal to the square root of the given number divides the number, than the number is composite; otherwise, it is prime.

Here are examples illustrating the "check":

57      $\sqrt{57} \approx 7.55$

Primes to check: 7, 5, 3, 2

57 is divisible by 3, so the number is composite.

589      $\sqrt{589} \approx 24.27$

Primes to check: 23, 19, 17, 13, 11, 7, 5, 3, 2

589 is divisible by 19, so the number is composite.

229      $\sqrt{229} \approx 15.13$

Primes to check: 13, 11, 7, 5, 3, 2

229 is not divisible by any of the numbers, so the number is prime.

1423 $\sqrt{1423} \approx 37.72$

Primes to check: 37, 31, 29, 23, 19, 17, 13, 111, 7, 5, 3, 2

1423 is not divisible by any of the numbers, so the number is prime.

Here are possible solutions for the first activity.

601	201
47	53
283	467
59	89

The last activity is more difficult. It requires students to create five different prime number using all ten number tiles. There are no conditions on the size of each number; however, the 0-tile cannot be placed in front of a number, like 07. Here are some possible solutions:

2	47	61	83	509
5	23	41	89	607
3	5	67	281	409

38 DEVELOPING ALGEBRAIC THINKING © Didax Educational Resources

PRIMES AND COMPOSITES

Teacher's Notes

Prime Stack *Prime Stack* involves using all 10 number tiles to create four prime numbers, a 1-, 2-, 3-, and 4-digit number. Some solutions are given below.

3	61	409	2857
5	47	601	8923
2	43	709	5861

This activity piggybacks what was done in *More Primes*. Creating five composite numbers, regardless of size, can be accomplished in many ways. In fact, six or seven composite numbers can be created: 4, 9, 27, 30, 51, 68; 4, 6, 8, 9, 27, 30, 51

LOTS OF PRIMES, LOTS OF COMPOSITES

Use any 8 number tiles 0 through 9 to create four 2-digit prime numbers.

Record your solution below:

_____

_____

_____

_____

Find a different solution and record below:

_____

_____

_____

_____

Now, use any 8 number tiles 0 through 9 to create four 2-digit composite numbers.

Record your solution below:

_____

_____

_____

_____

Find a different solution and record below:

_____

_____

_____

_____

PRIMES AND COMPOSITES

Use all 10 number tiles 0 through 9 to create three 2-digit prime numbers and two 2-digit composite numbers.

Record two solutions below:

___ ___   ___ ___

___ ___   ___ ___

___ ___

Now, use all 10 number tiles 0 through 9 to create three 2-digit composite numbers and two 2-digit prime numbers.

Record two solutions below:

___ ___   ___ ___

___ ___   ___ ___

___ ___

DEVELOPING ALGEBRAIC THINKING

MORE PRIMES

Use all 10 number tiles 0 through 9 to create two 3-digit prime numbers and two 2-digit prime numbers.

Record your solution below and find a different solution.

(i) ___ ___ ___

___ ___

___ ___ ___

___ ___

(ii) ___ ___ ___

___ ___

___ ___ ___

___ ___

Now, use all 10 number tiles to create 5 different primes. There are no conditions on the number of digits for the number; however the 0-tile cannot be placed in front of a number, like 07.

Record your results below.

(i) _____

(ii) _____

(iii) _____

(iv) _____

(v) _____

PRIME STACK

Use all 10 number tiles 0 through 9 to create four prime numbers that meet the conditions shown by the squares below.

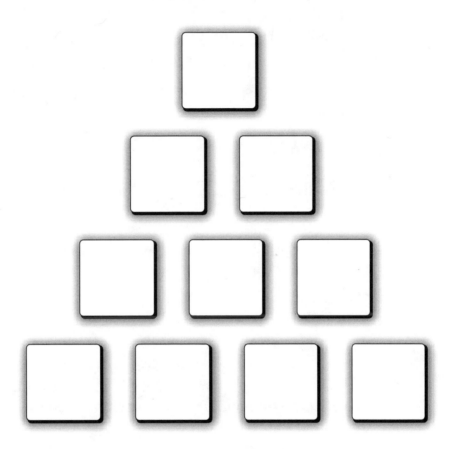

Now, use all 10 number tiles to create 4 different primes.

(i) _____

(ii) _____

(iii) _____

(iv) _____

ALL PRIMES

Use all 10 number tiles 0 through 9 to create five prime numbers. The number of digits in each number can be different. Record your results below.

(i) _____

(ii) _____

(iii) _____

(iv) _____

(v) _____

Find five different prime numbers. Record your results below.

(i) _____

(ii) _____

(iii) _____

(iv) _____

(v) _____

ALL COMPOSITES

Now, use all 10 number tiles to create five composite numbers. Again, the number of digits in each number can be different. Record your results below.

(i) _____

(ii) _____

(iii) _____

(iv) _____

(v) _____

Find five different composite numbers. Record your results below.

(i) _____

(ii) _____

(iii) _____

(iv) _____

(v) _____

DEVELOPING ALGEBRAIC THINKING

Divisibility Rules

DIVISIBILITY RULES

Teacher's Notes

Introduction We often need to know quickly whether one number will divide evenly, into another number. Knowing simple divisibility rules can make later factoring tasks in algebra much easier. The activities in this section provide opportunities to use the most common divisibility tests to create numbers satisfying the rules. Starting with tasks involving a limited number of tiles and divisibility rules, the pages become increasingly more difficult until all 10 number tiles are used.

Looking at the Algebra A number is said to be divisible by another number if and only if the first number divides evenly into the second number, leaving a remainder of 0. Divisibility tests are helpful in algebra when students begin factoring various polynomials.

Below is a list of the most common divisibility rules.

A number is divisible by …

- 2, if it ends with an even number in the ones place. That is, the ones digit is 0, 2, 4, 6, or 8.

  Examples: 5<u>4</u>, 80<u>6</u>, 379<u>8</u>.

- 3, if the sum of the digits in the number is divisible by 3.

  Examples: 57, 321, 5412.

- 4, if the last two-digit number is divisible by 4.

  Examples: 1<u>24</u>, 63<u>80</u>, 9999<u>12</u>.

- 5, if the ones digit is 0 or 5.

  Examples: 75, 200, 4070, 99995

- 6, if the number is divisible by 2 and by 3.

  Examples: 48, 234, 7236.

- 8, if the last three-digit number is divisible by 8.

  Examples: 1<u>240</u>, 3<u>568</u>, 765<u>032</u>.

- 9, if the sum of the digits in the number is divisible by 9.

  Example: 468, 5247, 71559.

- 10, if the number ends in 0.

  Example: 5<u>0</u>, 70<u>0</u>, 408<u>0</u>.

Tests for 7, 11, and other numbers do exist, but they are sometimes very awkward to use. Long division or using a calculator are often quicker.

© Didax Educational Resources · **DEVELOPING ALGEBRAIC THINKING**

DIVISIBILITY RULES

Teacher's Notes

Divisibility Rules 1, 2, 3, and 4

For *Divisibility Rules 1, 2, 3,* and *4* only 6 number tiles are used for each task and only two divisibility rules are involved. Many solutions exist for each task. In creating these numbers, students must consider place value ideas, along with divisibility tests. The least number would have the 1-digit in the hundreds place, most likely followed by 0 in the tens place. The greatest number would have a 9 in the hundreds place, possibly followed by 8 in the tens place. Below are the least and greatest numbers satisfying the divisibility rules.

	Least	Greatest
Divisibility Rules 1 (2 and 3)	102	984
Divisibility Rules 2 (3 and 5)	105	975
Divisibility Rules 3 (3 and 4)	108	984
Divisibility Rules 4 (4 and 9)	108	972

Divisibility Rules 5 and 6

In *Divisibility Rules 5* and *Divisibility Rules 6*, nine number tiles are used to create three 3-digit numbers satisfying the listed divisibility tests. The question: "Where should I start?" brings problem solving into play. Divisibility by 2 requires an even digit in the ones place and divisibility by 4 requires the last 2-digit number to be divisible by 4. In each task, these conditions provide what appears to be the logical "starting place." Having met those conditions, students have many possibilities for creating numbers divisible by 3.

Divisibility Rules 7, 8, 9, 10, 11, and 12

Divisibility Rules 7, 8, 9, 10, 11, and *12* each use all 10 numbers tiles to create two 5-digit numbers satisfying two divisibility rules. Again, the nature of a particular test essentially dictates a starting place, unless a strategy of "Guess and check" is used.

Divisibility Rules 13, 14, and 15

Divisibility Rules 13, 14, and *15* are the most difficult activities in this section. Each requires students to use all 10 number tiles to create several multiplace numbers satisfying a variety of divisibility rules. Certain digits are strategically placed to provide students with limited choices for particular digits. For all three tasks, students cannot merely start on the first number and work through the numbers in order.

DIVISIBILITY RULES

Teacher's Notes

For *Divisibility Rules 13*, here is a list of possible steps to solve the task.

- For ☐342 to be divisible by 9, the only choice of a digit is 9.
- For 632☐ to be divisible by 6 (2 and 3), the only choice of a digit is 4.
- For 4☐2☐ to be divisible by 5 and 8, the ones digit must be 0.
- For 9☐7☐ to be divisible by 4 and 9, the ones digit could be 2 or 6. However, with 2, there is no digit remaining that will satisfy divisibility by 9. So, the ones digit is 6 and the hundreds digit is 5.
- For 1☐2☐ to be divisible by 3 and 4, only 8 remains to satisfy divisibility by 4.
  Therefore, the hundreds digit could be 1 or 7. Try 1 first.
- For 3☐☐6 to be divisible by 4 and 6, the only remaining choices are 7 in the tens place and 2 in the hundreds place.
- This leaves 3 in the hundreds place for 4☐2☐.
- If 7 is used as the hundreds digit in the first number 1☐2☐, then for 3☐☐6 to be divisible by 4 and 6, the only remaining choices are 1 in the tens place and 2 in the hundreds place.
- This also leaves 3 in the hundreds place for 4☐2☐.

For *Divisibility Rules 14*, here is a list of possible steps to solve the task.

- For 910☐ to be divisible by 3 and 5, the only choice of a digit in the ones place is 5.
- For 10☐5 to be divisible by 3, 5, and 9, the only choice of a digit in the tens place is 3.
- For 35☐ to be divisible by 6, the only choice of a digit in the ones place is 4.
- For 75☐☐ to be divisible by 4, 6, 9, and 10, the only choice of a digit in the ones place is 0. Now, the only choice for the tens place is 6.
- For 472☐ to be divisible by 3 and 4, the only choice of a digit in the ones place is 8.
- For 61☐☐ to be divisible by 4 and 9, the only choice of a digit in the ones place is 2. Now, the only choice for the tens place is 9.
- This leaves 1 and 7 for 1☐☐. The digits can be placed in either order 117 or 171.

© Didax Educational Resources   **DEVELOPING ALGEBRAIC THINKING**

DIVISIBILITY RULES

Teacher's Notes

Divisibility Rules 15

Divisibility Rules 15 has all 3-digit numbers. Here is a list of possible steps to solve this task.

- For 9☐6 to be divisible by 4 and 9, the only choice of a digit in the tens place is 3.

- For ☐7☐ to be divisible by 2 and 5, the only choice of a digit in the ones place is 0.

- For 9☐☐ to be divisible by 3 and 5, the only choice of a digit in the ones place is 5.

- For 64☐ to be divisible by 3 and 4, the only choice of a digit in the ones place is 8.

- For ☐0☐ to be divisible by 2 and 9, the only choice of a digit in the ones place is 2, and therefore, a 7 in the hundreds place.

- Now, for 3☐☐ to be divisible by 2 and 3, the only choice of a digit in the ones place is 6, and therefore, the tens digit is 9.

- For 9☐☐, the tens digit is 1 or 4, and for ☐7☐, the hundreds digit is 4 or 1.

DIVISIBILITY RULES

Divisibility Rules 1 Use any 6 number tiles 0 through 9 to create two 3-digit numbers, each divisible by 2 and 3.

Find the smallest 3-digit number divisible by 2 and 3. Find the largest.

Smallest   Largest

Divisibility Rules 2 Use any 6 number tiles 0 through 9 to create two 3-digit numbers, each divisible by 3 and 5.

Find the smallest 3-digit number divisible by 3 and 5. Find the largest.

Smallest   Largest

DIVISIBILITY RULES

Divisibility Rules 3 Use any 6 number tiles 0 through 9 to create two 3-digit numbers, each divisible by 3 and 4.

Find the smallest 3-digit number divisible by 3 and 4. Find the largest.

Smallest   Largest

Divisibility Rules 4 Use any 6 number tiles 0 through 9 to create two 3-digit numbers, each divisible by 4 and 9.

Find the smallest 3-digit number divisible by 4 and 9. Find the largest.

Smallest   Largest

DEVELOPING ALGEBRAIC THINKING — World Teacher Press®

DIVISIBILITY RULES

Divisibility Rules 5 Use any 9 number tiles 0 through 9 to create three 3-digit numbers, each divisible by 2 and 3.

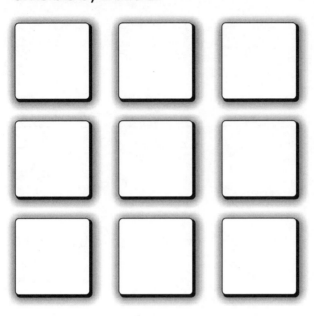

Now, find three different numbers, each divisible by 2 and 3.

Divisibility Rules 6 Use any 9 number tiles 0 through 9 to create three 3-digit numbers, each divisible by 3 and 4.

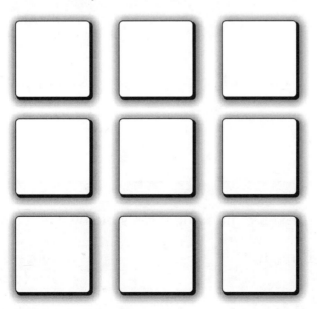

Now, find three different numbers, each divisible by 3 and 4.

DIVISIBILITY RULES

Divisibility Rules 7 Use all 10 number tiles 0 through 9 to create two 5-digit numbers, each divisible by 3 and 5.

Find the smallest 5-digit number divisible by 3 and 5. Find the largest.

Smallest   Largest

Divisibility Rules 8 Use all 10 number tiles 0 through 9 to create two 5-digit numbers, each divisible by 4 and 6.

Find the smallest 5-digit number divisible by 4 and 6. Find the largest.

Smallest   Largest

DIVISIBILITY RULES

Divisibility Rules 9 Use all 10 number tiles 0 through 9 to create two 5-digit numbers, each divisible by 2 and 9.

Find the smallest 5-digit number divisible by 2 and 9. Find the largest.

Smallest   Largest

Divisibility Rules 10 Use all 10 number tiles 0 through 9 to create two 5-digit numbers, each divisible by 4 and 9.

Find the smallest 5-digit number divisible by 4 and 9. Find the largest.

Smallest   Largest

DEVELOPING ALGEBRAIC THINKING

DIVISIBILITY RULES

Divisibility Rules 11 Use all 10 number tiles 0 through 9 to create two 5-digit numbers, each divisible by 2 and 3.

Find the smallest 5-digit number divisible by 2 and 3. Find the largest.

Smallest   Largest

Divisibility Rules 12 Use all 10 number tiles 0 through 9 to create two 5-digit numbers, each divisible by 5 and 9.

Find the smallest 5-digit number divisible by 5 and 9. Find the largest.

Smallest   Largest

DIVISIBILITY RULES

Divisibility Rules 13 Use all 10 number tiles 0 through 9 to create six 4-digit numbers, each divisible by the given numbers.

Divisible by 3 and 4 | 1 | | 2 | |

Divisible by 6 | 6 | 3 | 2 | |

Divisible by 4 and 6 | 3 | | | 6 |

Divisible by 5 and 8 | 4 | | 2 | |

Divisible by 9 | | 3 | 4 | 2 |

Divisible by 4 and 9 | 9 | | 7 | |

© Didax Educational Resources — DEVELOPING ALGEBRAIC THINKING

DIVISIBILITY RULES

Divisibility Rules 14 Use all 10 number tiles 0 through 9 to create seven 3-digit or 4-digit numbers, each divisible by the given numbers.

Divisible by 3 and 5: **9 1 0** ☐

Divisible by 3, 5, and 9: **1 0** ☐ **5**

Divisible by 4 and 9: **6 1** ☐ ☐

Divisible by 3 and 4: **4 7 2** ☐

Divisible by 6: **3 5** ☐

Divisible by 3 and 9: **1** ☐ ☐

Divisible by 4, 6, 9, and 10: **7 5** ☐ ☐

DIVISIBILITY RULES

Divisibility Rules 15 Use all 10 number tiles 0 through 9 to create six 3-digit numbers, each divisible by the given numbers.

Divisible by 2 and 9 ☐ 0 ☐

Divisible by 2 and 5 ☐ 7 ☐

Divisible by 3 and 4 6 4 ☐

Divisible by 3 and 5 9 ☐ ☐

Divisible by 2 and 3 3 ☐ ☐

Divisible by 4 and 9 9 ☐ 6

TEACHER'S NOTES

Locker Numbers

LOCKER NUMBERS

Teacher's Notes

Introduction The Locker Number problems are designed to develop number sense, a necessary prerequisite for algebraic thinking. "Guess and check" is certainly one of the main problem solving strategies to use in finding solutions to locker problems with number tiles. Additionally, showing possible solutions with the tiles and then writing equations are strategies used with these problems. In the discussion that follows, both a "mathematical reasoning" solution and an algebraic solution are provided for Problem 1, and thereafter, only an algebraic solution. *Locker Numbers 1* has unique locker numbers, while *Locker Numbers 2* has more than one locker number for some problems. The first digit refers to the digit in the hundreds place, whereas the third digit refers to the digit in the ones place.

Looking at the Algebra

Locker Numbers 1

Problem 1 In Problem 1, the ones digit (or third digit) is 0, 2, 4, 6, or 8, since the number is even. The third digit is 3 times the first digit. There is only one possibility for the first digit and the third digit. Since 0 x 3 = 0, a second 0 tile would be needed. Multiplying 4, 6, or 8 by 3 produces a 2-digit product. Therefore, the first digit must be 2 and the third digit is 6. The second digit is 3 more than the first, so 2 + 3 = 5. The locker number is 256. Algebraically, if the locker number is xyz, then:

$x + y + z < 15$ $z = 3x$ $y = x + 3$

Substituting gives $x + y + 3 + 3x < 15$, or $5x < 12$, or $x < 12/5$.

So x = 0, 1, or 2.

With x = 0, z would be 0.

With x = 1, the locker number would be 143, which is odd.

So, x = 2, y = 5, and z = 6.

The locker number is 256.

Problem 2 For Problem 2, if the locker number is xyz, then:

$y = z - 7$ $x = y + 2 = z - 7 + 2 = z + 5$ $z = 0, 2, 4, 6,$ or 8

With z = 0, 2, 4, or 6, y is a negative integer.

So z = 8, y = 1, and x = 3.

The locker number is 318.

LOCKER NUMBERS

Teacher's Notes

Problem 3 For Problem 3, if the locker number is xyz, then:

$x + y + z = 7$ $\qquad$ $x = 2y$ $\qquad$ $x, y, z \neq 0$

So, $2y + y + z = 7$, or $3y + z = 7$.

Since the sum is 7, $y = 1$ or 2; otherwise, the sum would be greater than 7.

If $y = 1$, then $x = 2$ and $z = 4$.

The locker number would be 214, which is even.

Therefore, $y = 2$, $x = 4$, and $z = 1$.

The locker number is 421.

Problem 4 For Problem 4, if the locker number is xyz, then:

$xyz > 200$

So $x = 2, 3, 4, 5, 6, 7, 8,$ or 9 $\qquad$ $y = 3x$

So $y = 6$ or 9 and $x = 2$ or 3

$z = y - 4 = 3x - 4$

So $z = 2$ or 5

Since $x \neq z$ (there is only one 2-tile), the locker number must be 395.

Problem 5 For Problem 5, if the locker number is xyz, then:

$x + y + z > 15$ $\qquad$ $x = z + 1$ $\qquad$ $y = x + 2 = z + 1 + 2 = z + 3$

So $z + 1 + z + 3 + z > 15$, or $3z + 4 > 15$, or $z > 3\ 2/3$

Since the locker number is odd, $z = 5, 7,$ or 9.

If $z = 7$, then y is a 2-digit number.

If $z = 9$, then both x and y are 2-digit numbers.

Therefore, $z = 5$, $x = 6$, and $y = 8$.

The locker number is 685.

Problem 6 For Problem 6, if the locker number is xyz, then:

$y = z - 4$ $\qquad$ $x = y + 7$

So $x = z - 4 + 7 = z + 3$

Since the locker number is odd, $z = 1, 3, 5, 7,$ or 9.

With $z = 1$ or 3, y is a negative integer.

With $z = 7$ or 9, x is a 2-digit number.

So $z = 5$, $y = 1$, and $x = 8$.

The locker number is 815.

LOCKER NUMBERS

Teacher's Notes

Locker Numbers 2

Problem 1   For Problem 1, if the locker number is xyz, then:
$y = 2z$ $x = 2y = 2(2z) = 4z$
Since $z \neq 0$ (otherwise x, y = 0), $z = 1$ or 2.
Other values of z will make x or y a 2-digit number.
If $z = 1$, then $y = 2$, and $x = 4$.
The locker number is 421.
If $z = 2$, then $y = 4$ and $x = 8$.
The locker number is 842.

Problem 2   For Problem 2, if the locker number is xyz, then:
$y = x + z$
With xyz > 700, $x = 7$ or 8. If $x = 7$, then $z = 1$ and $y = 8$, or $z = 2$ and $y = 9$.
The locker numbers are 781 or 792.
If $x = 8$, then $z = 1$ and $y = 9$.
The locker number is 891.

Problem 3   For Problem 3, if the locker number is xyz, then:
$z = 3, 6,$ or 9 $y = x - 6$
Since y must be non-negative, the only possible values for x are 6, 7, 8, or 9.
If $x = 9$, then $y = 3$ and $z = 6$.
If $x = 8$, then $y = 2$ and $z = 6$.
If $x = 7$, then $y = 1$ and $z = 6$.
If $x = 6$, then $z = 6$.
The locker numbers are 936, 826, and 716.

Problem 4   For Problem 4, if the locker number is xyz, then:
$x = y + 3$ $y = z + 3$ so $x = y + 3 = z + 3 + 3 = z + 6$
Since xyz is odd, $z = 1$ or 3.
If $z = 1$, then $y = 4$ and $x = 7$.
The locker number is 741.
If $z = 3$, then $y = 6$ and $x = 9$.
The locker number is 963.

DEVELOPING ALGEBRAIC THINKING

LOCKER NUMBERS

Teacher's Notes

Problem 5 For Problem 5, if the locker number is xyz, then:

xyz < 500   z = y − 2

Since 2 is a factor of x, x = 2 or 4.

If x = 2, then y = 6 and z = 4, or y = 8 and z = 6.

The locker numbers are 264 and 286.

If x = 4, then y = 2 and z = 0, or y = 8 and z = 6.

The locker numbers are 420 and 486.

Problem 6 For Problem 6, if the locker number is xyz, then:

z = 0, 2, 4, 6, or 8   x = z − 1   y = z + 4

If z = 0, then x is a negative integer.

If z = 6 or 8, then y is a 2-digit number.

So, z = 2 or 4. If z = 2, then x = 1 and y = 6.

If z = 4, then x = 3 and y = 8.

The locker numbers are 162 and 384.

DEVELOPING ALGEBRAIC THINKING

LOCKER NUMBERS 1

Use the number tiles 0 through 9 to find the 3 digit locker numbers.

1. The sum of the digits is less than 15.
   The third digit is 3 times the first.
   The second digit is 3 more than the first.
   The number is even.

   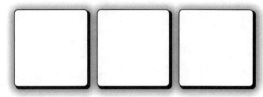

2. The first digit is 2 more than the second digit.
   The second digit is 7 less than the third digit.
   The number is even.

3. The sum of the digits is 7.
   No digit is 0.
   The first digit is twice the second digit.
   The number is odd.

4. The second digit is triple the first digit.
   The locker number is greater than 200.
   The third digit is 4 less than the second digit.

5. The locker number is odd.
   The sum of the digits is greater than 15. The first digit is 1 more than the third.
   The second digit is 2 more than the first.

6. The locker number is odd.
   The second digit is 4 less than the third.
   The first digit is 7 more than the second.

   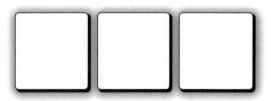

LOCKER NUMBERS 2

Use the number tiles 0 through 9 to find the 3 digit locker numbers.

1. The first digit is twice the second digit.

   The second digit is twice the third digit.

2. The second digit is the sum of the first and third digits.

   The number is greater than 700.

3. The third digit is a positive multiple of 3.

   The second digit is 6 less than the first digit.

   The sum of the digits is even.

4. The first digit is 3 more than the second digit.

   The second digit is 3 more than the third digit.

   The number is odd.

5. Two is a factor of all 3 digits.

   The number is less than 500.

   The third digit is 2 less than the second digit.

6. The locker number is even.

   The second digit is 4 more than the third.

   The first digit is 1 less than the third.

DEVELOPING ALGEBRAIC THINKING

Exciting Exponents

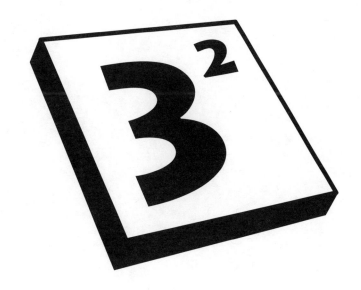

EXCITING EXPONENTS

Teacher's Notes

Introduction A common topic in the algebra curriculum is that of exponents. Later, as work with the distance formula begins, students observe sums of squares as the radicand (the number beneath the radical). Proficiency with number theory concepts, for example squaring or cubing a number, often makes work much quicker. Estimation strategies also aid in making decisions.

Looking at the Algebra

Exciting Exponents 1 In *Exciting Exponents 1*, basic rules of exponents with whole numbers are covered, along with simple number theory ideas.

- Any positive whole number raised to the zero power is 1.
- 1 raised to any whole number power is 1.
- 5 raised to any positive whole number power always has a 5 in the ones place.

Using these three ideas and some estimation strategies, the 10 tiles can be placed to satisfy the five equations. There are two distinct solutions.

$$\boxed{7}^{\boxed{0}} = 1 \qquad \boxed{1}^{\boxed{9}} = 1 \quad \text{or} \quad \boxed{9}^{\boxed{0}} = 1 \qquad \boxed{1}^{\boxed{7}} = 1$$

$$2^{\boxed{3}} = \boxed{8}$$

$$\boxed{5}^{2} = 2\boxed{5}$$

$$2^{\boxed{6}} = 6\boxed{4}$$

Exciting Exponents 2 *Exciting Exponents 2* continues with the ideas above, but also focuses on the relative magnitude of numbers raised to various powers.

$$\boxed{5}^{3} = 1\boxed{2}5$$

$$2^{\boxed{4}} = 1\boxed{6}$$

$$7^{\boxed{0}} = \boxed{1}$$

$$\boxed{8}^{\boxed{3}} = 512$$

$$1\boxed{7}^{2} = 28\boxed{9}$$

EXCITING EXPONENTS

Teacher's Notes

Exciting Exponents 3

Exciting Exponents 3 is the first activity involving sums of powers of numbers. Again, the magnitude of the numbers plays a key role in deciding the location of the ten tiles. The first equation illustrates a basic Pythagorean Triple (6, 8, 10). In the second equation, the resulting sum of 7 limits the exponents to 2 and 1. Similarly, in the last equation, 41 limits the base numbers to 4 and 5. The sum of 126 is 125 + 1, where 125 is the perfect cube 5^3. The exponent of 4 must by 0. This leaves 7 and 9 to complete the third equation.

Exciting Exponents 4

The last two activities in the section focus on sums of squares. There are several starting points for *Exciting Exponents 4*. Below is one possibility.

- In the fourth equation, 5^2 and a sum of thirty-something limits the equation to $5^2 + 3^2 = 34$.

- Quick subtraction in the second equation shows that, with a sum of ninety-something and with subtracting 3^2, the missing digit to be squared must give a result in the eighties. Only 9^2 works; therefore, the missing digit in the sum is 0.

- To have a sum that ends in 5 for the third equation requires that the number to be squared must end in a 6 (49 + _6). Two possibilities exist, 4 and 6; however, the 4-digit has been used. So, we have $7^2 + 6^2 = 85$.

- The first equation must have 7^2 to get close to 53. Therefore, the second base number must be 2.

- This leaves $2^2 + 1^2 = 5$.

Exciting Exponents 5

Exciting Exponents 5 is more difficult than *Exciting Exponents 4* due to the magnitude of the numbers involved.

- In the third equation, since the sum is 3 digits in length, the choices for the second addend are limited to 6^2, 7^2, 8^2, or 9^2. However, only $8^2 + 6^2$ produces a perfect square sum, 10^2.

- With a 5 in the ones place of the sum in the second equation, the first addend is limited to 3^2 or 7^2, which would give sums of 25 or 65, respectively. Since the 6-tile was used in the third equation, the solution here must be $3^2 + 4^2 = 25$.

- In the fourth equation $25^2 + 625$. There are two pairs of digits remaining that, when squared, produce a 5 as the ones digit, 8^2 and 9^2 (_4 + _1) or 4^2 and 7^2 (_6 + _9). However, 28^2 or 29^2 are both greater than 25^2. The solution must be $24^2 + 7^2 = 25^2$.

- This leaves $8^2 + 15^2 = 289$ for the first equation.

EXCITING EXPONENTS 1

Place all 10 number tiles 0 through 9 in the squares to make correct number sentences.

$\square^{\square} = 1$

$1^{\square} = \square$

$2^{\square} = \square$

$\square^{2} = \square 5$

$2^{\square} = 6\square$

EXCITING EXPONENTS 2

Place all 10 number tiles 0 through 9 in the squares to make correct number sentences.

$\square^3 = 1\square5$

$2^\square = 1\square$

$7^\square = \square$

$\square^\square = 512$

$1\square^2 = 28\square$

EXCITING EXPONENTS 3

Place all 10 number tiles 0 through 9 in the squares to make correct number sentences.

☐² + ☐² = 1 0 0

2☐ + 3☐ = 7

☐² + ☐² = 1 3 0

5☐ + 4☐ = 1 2 6

☐² + ☐² = 4 1

EXCITING EXPONENTS 4

Place all 10 number tiles 0 through 9 in the squares to make correct number sentences.

$\square^2 + \square^2 = \boxed{5}\boxed{3}$

$\square^2 + \boxed{3}^2 = \boxed{9}\square$

$\boxed{7}^2 + \square^2 = \square\boxed{5}$

$\boxed{5}^2 + \square^2 = \boxed{3}\square$

$\boxed{2}^2 + \square^2 = \square$

EXCITING EXPONENTS 5

Place all 10 number tiles 0 through 9 in the squares to make correct number sentences.

$$\square^2 + 1\square^2 = 2\,8\,\square$$

$$\square^2 + 4^2 = \square\,5$$

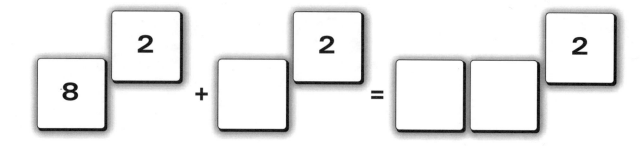

$$8^2 + \square^2 = \overline{\square\square}^{\,2}$$

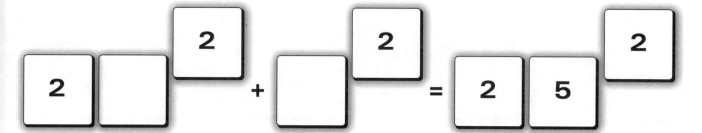

$$2\square^2 + \square^2 = 2\,5^2$$

TEACHER'S NOTES

Plus and Times

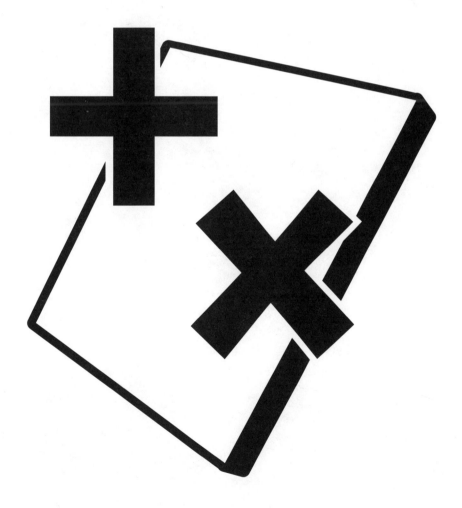

PLUS AND TIMES

Teacher's Notes

Introduction Two operation signs, + and x, provide three easy number tile activities that give an introduction to "Developing Algebraic Thinking." Number sense is important. As students begin to look for patterns in various solutions, a need for algebra becomes clear in order to analyze each activity.

Looking at the Algebra

Plus 5 In *Plus 5*, numerous solutions exist. If we label the 5 digits as shown in the diagram below, then we have the equation **b + a + c = d + a + e**

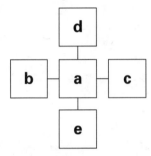

Simplifying, we obtain the equation **b + c = d + e**. In other words, the algebra shows that we only need to find two pairs of number tiles that have the same sum. The number tile **a** can be any digit 0 through 9.

The least possible sum occurs when a = 0.

The smallest possible sum satisfying the equation

b + c = d + e is 5 = 1 + 4 = 2 + 3.

The greatest possible sum occurs when a = 9.

That sum is 22 and it occurs when b + c = d + e = 13.

So 13 = 5 + 8 = 6 + 7.

DEVELOPING ALGEBRAIC THINKING

PLUS AND TIMES

Teacher's Notes

Plus 9 *Plus 9* also has numerous solutions. The algebra is very similar to what is described for Plus 5. Look at the diagram below.

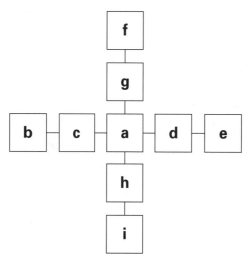

In this problem, we have the equation $b + c + a + d + e = f + g + a + h + i$.

This simplifies to $b + c + d + e = f + g + h + i$.

Once again, the number in the center can be any digit 0 through 9. We need two groups of four digits that have the same sum. Lots of possibilities exist; however, an additional method to solve the problem is to create two Plus 5s as shown below.

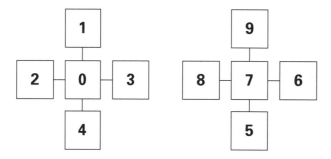

Now, move a vertical pair from one plus to the other and a horizontal pair from one plus to the other. The problem is solved.

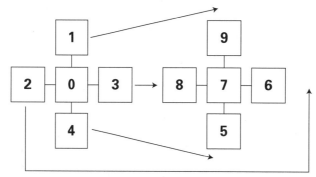

PLUS AND TIMES

Teacher's Notes

Times In *Times*, numerous solutions also exist. If we label the 5 digits as before, then we have the equation b x a x c = d x a x e.

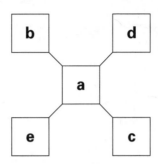

Simplifying, we obtain the equation b x c = d x e if a ≠ 0

The algebra shows that we need to find two pairs of number tiles that have the same product. The number tile a can be any digit 1 through 9. If a, however, is 0, it does not matter whether or not the products are the same since the final product will still be 0.

The greatest possible product of 216 occurs when a = 9, and the product of each pair is 24 = 3 x 8 = 4 x 6.

Can all the digits 0 through 9 be used for b, c, d, and e?

No! Certainly 0 is out, since the product would be 0. Since there are no multiples of 5 and 7, these two tiles cannot be used either. All three, however, will work as values for a.

© Didax Educational Resources   **DEVELOPING ALGEBRAIC THINKING**   **83**

PLUS 5

Place any 5 of the digits 0 through 9 in the squares so that the sum of each line is the same.

Five number tiles have been used in the Plus 5 above. Five number tiles remain. Try to use those five tiles to create another Plus 5.

What is the least possible sum you can find for Plus 5? Record your results in the Plus 5 provided.

What is the greatest possible sum you can find Plus 5? Record your results in the Plus 5 provided.

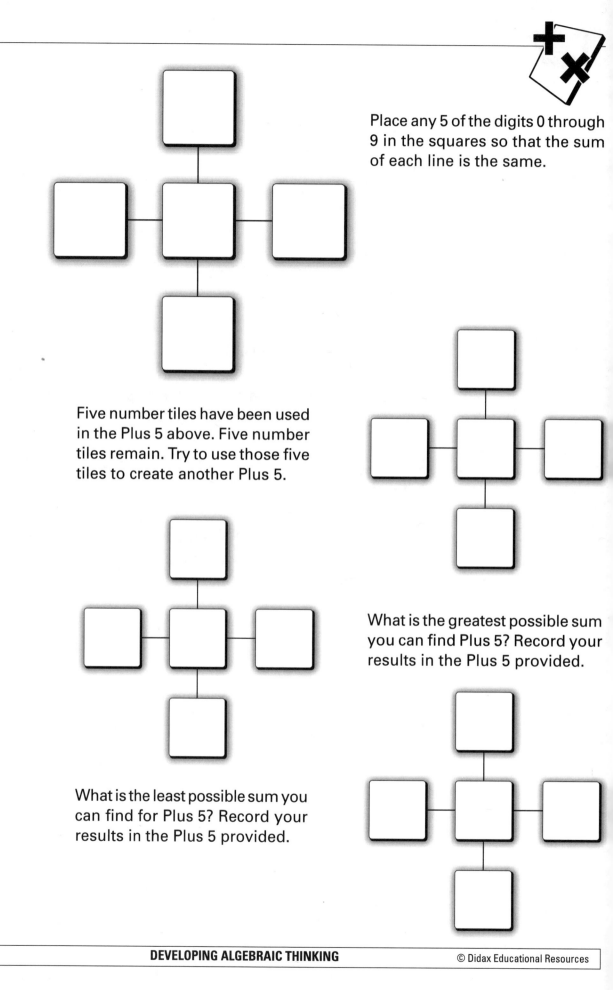

PLUS 9

Place any 9 of the digits 0 through 9 in the squares so that the sum of each line is the same.

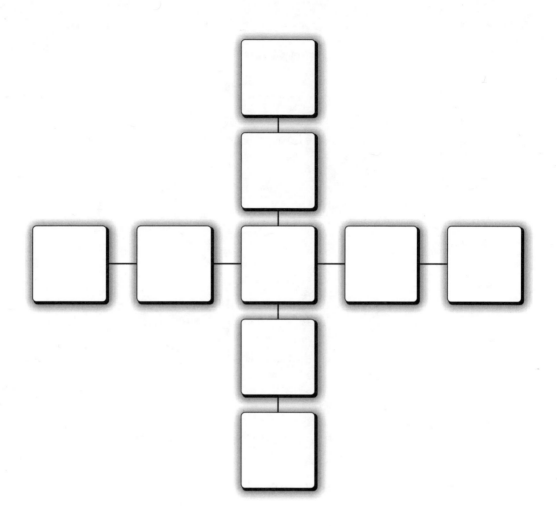

What is the greatest possible sum you can find? _____

What is the least possible sum you can find? _____

TIMES

Place any 5 of the digits 0 through 9 in the squares so that the product of each line is the same.

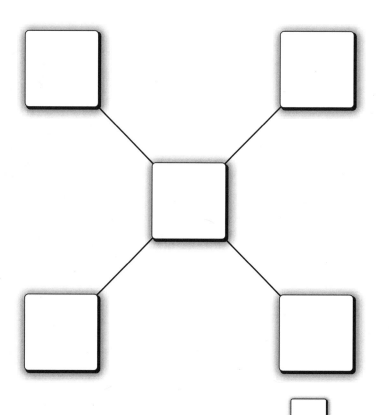

What is the greatest possible product you can find?

Record your result in the Times provided.

What is the least possible product you can find?

Record your result in the Times provided.

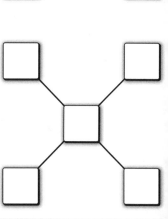

Shapes

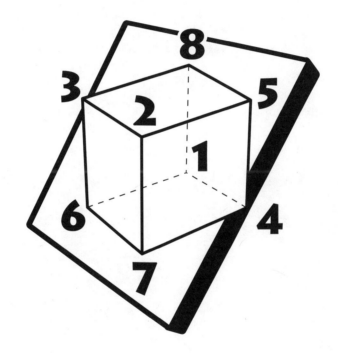

SHAPES

Teacher's Notes

Introduction Shapes begins with a small triangle using only six tiles and ends with Box Sum, a two-dimensional representation of a cube using eight tiles. The equations become progressively more difficult to solve simultaneously; however, several patterns do appear.

Looking at the Algebra

Triangle 6 In Triangle 6, the algebra is straightforward. Each of the three digits a, c, and e has to be used in two equations as shown below.

```
      a
   b     f
  c   d   e
```

$a + b + c = c + d + e = e + f + a$

$a + b = d + e, \quad c + d = f + a$

To create the smallest sum 6, we use 0, 1, and 2 at the vertices; for the largest sum 21 we use 7, 8, and 9.

Notice the clockwise order of the digits in the examples.

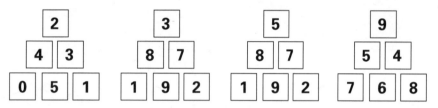

Triangle 9 For Triangle 9, the three digits in the vertices are again important in finding a solution.

```
        a
     b     i
    c       h
   d  e  f  g
```

$a + b + c + d = d + e + f + g = g + h + i + a$

$a + b + c = e + f + g, \quad d + e + f = a + i + h$

Notice the clockwise (or counterclockwise) order of consecutive digits in the following solutions.

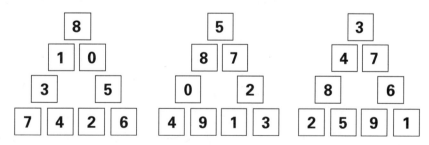

DEVELOPING ALGEBRAIC THINKING

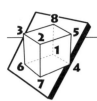

SHAPES

Teacher's Notes

Triangle 10   *Triangle 10* can be considered as an extension of *Triangle 9*, where the middle digit f is 0. Other solutions do exist, however, where f is not 0.

```
        a
      b   j
     c     i
    d e f g h
```

$a + b + c + d = d + e + f + g + h = h + i + j + a$
$a + b + c = e + f + g + h, \quad d + e + f + g = a + j + i$

In the solutions below, again notice the clockwise (or counter-clockwise) order of consecutive digits.

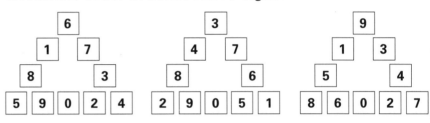

Here are solutions where f ≠ 0.

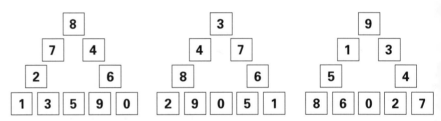

Rectangle   *Rectangle* is one of the most difficult number tile puzzles in the book. The equations are easily determined; however, finding the proper location of digits is not easily accomplished.

```
   a  j  i
   b     h
   c     g
   d  e  f
```

$a + b + c + d = d + e + f = f + g + h + i = i + j + a$
$a + b + c = e + f, \quad d + e = g + h + i, \quad f + g + h = j + a$

8	4	5		9	6	1
2		3		0		8
6		0		2		3
1	7	9		5	7	4

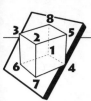

SHAPES

Teacher's Notes

Pentagon The equations for *Pentagon* can be solved so that the remaining equations each have a sum of two digits on both sides of the equal sign.

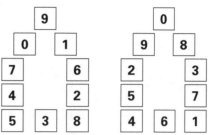

$a + b + c = c + d + e = e + f + g = g + h + i = i + j + a$

$a + b = d + e,$ $c + d = f + g,$ $e + f = h + i$

$g + h = j + a$ $b + c = i + j$

Box Sum The final problem in this section is *Box Sum*. It has appeared in many different puzzle books, usually with no mathematical background information. Since the sum of each face of the cube must be the same, there are six algebraic expressions that are equal. Subtracting like terms from various equations provides several relationships among the digits.

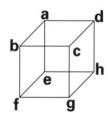

$a + b + c + d = a + b + f + e = a + d + h + e =$
$d + h + g + c = b + c + g + f = e + f + g + h$

$c + d = f + e,$ $b + f = d + h,$ $a + e = c + g$

$b + c = e + h,$ $a + b = g + h,$ $a + d = f + g$

All six of the equations show that the sum of the two digits on one edge of the cube is the same as the sum of the two digits on the edge diagonally across on the opposite face.

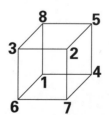 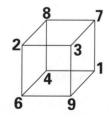 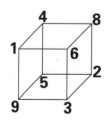

To start, select an edge and two digits. Now, select two other digits that give the same sum. With these initial selections, the remaining tiles become easier to place.

SHAPES

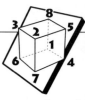

Triangle 6 Place any 6 of the digits 0 through 9 in the squares so that the sum of each line is the same.

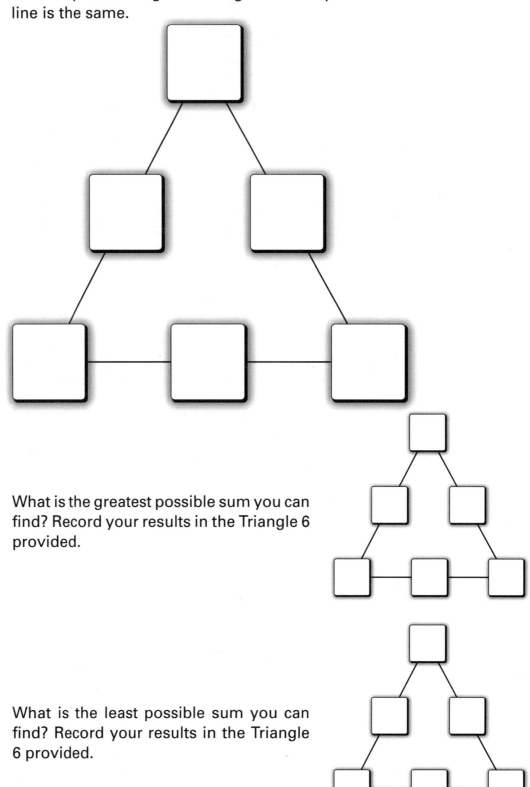

What is the greatest possible sum you can find? Record your results in the Triangle 6 provided.

What is the least possible sum you can find? Record your results in the Triangle 6 provided.

92 DEVELOPING ALGEBRAIC THINKING © Didax Educational Resources

SHAPES

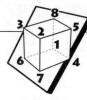

Triangle 9 Place any 9 of the digits 0 through 9 in the squares so that the sum of each line is the same.

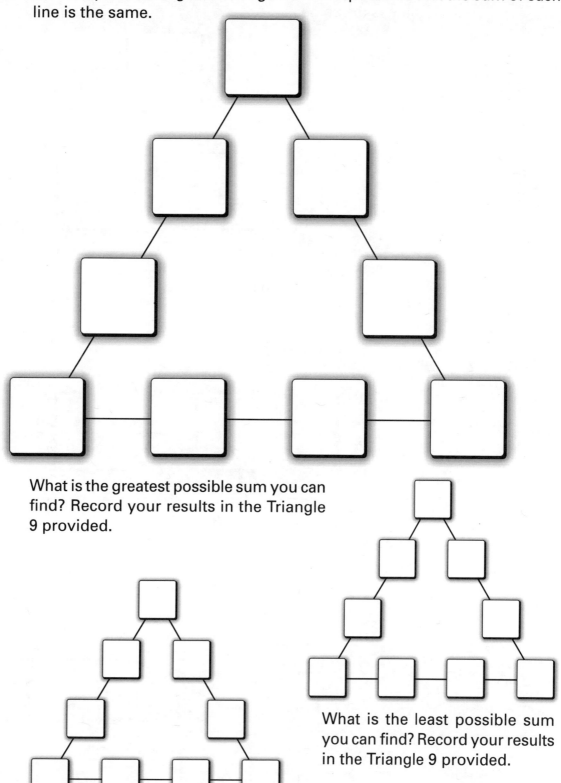

What is the greatest possible sum you can find? Record your results in the Triangle 9 provided.

What is the least possible sum you can find? Record your results in the Triangle 9 provided.

DEVELOPING ALGEBRAIC THINKING

SHAPES

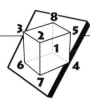

Triangle 10   Place all 10 of the digits 0 through 9 in the squares so that the sum of each line is the same.

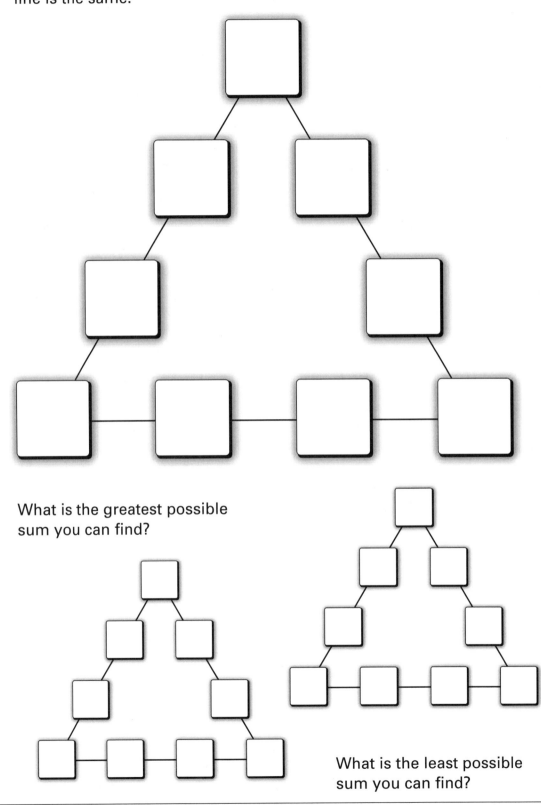

What is the greatest possible sum you can find?

What is the least possible sum you can find?

SHAPES

Rectangle Use all 10 number tiles 0 through 9 in the squares so that the sum of each line is the same.

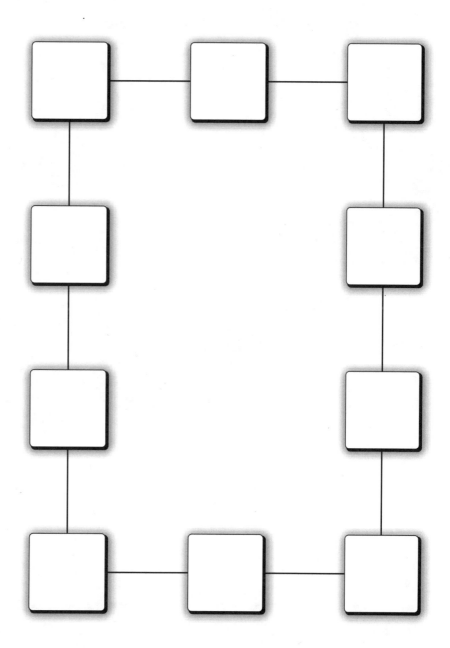

SHAPES

Pentagon   Use all 10 number tiles 0 through 9 in the square so that the sum of each line is the same.

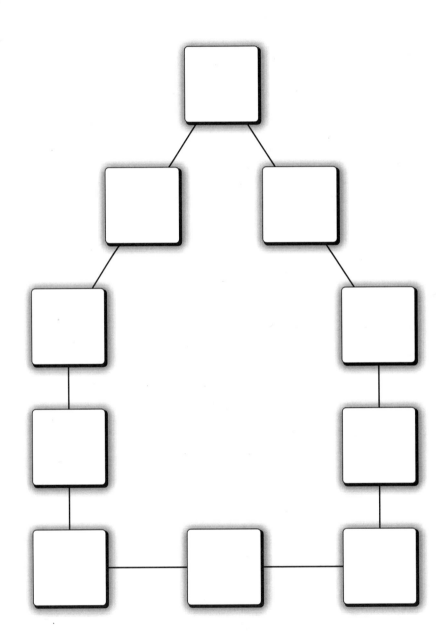

SHAPES

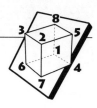

Box Sum Place any 8 of the number tiles 0 through 9 in the squares so that the sum of the digits on every face of the box is the same.

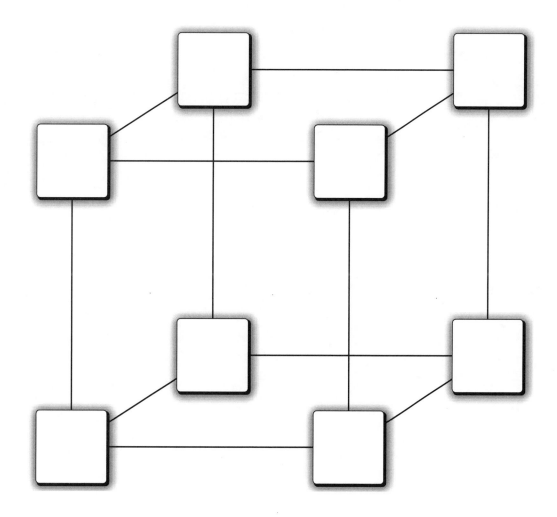

TEACHER'S NOTES

Pentomino Puzzles

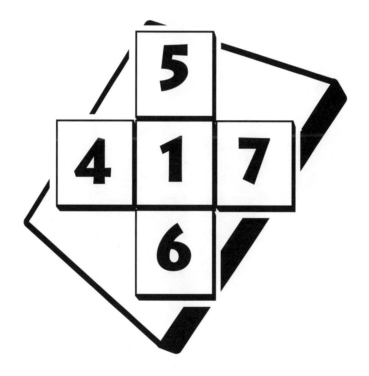

PENTOMINO PUZZLES

Teacher's Notes

Introduction Geometrically, a polyomino is a set of squares connected side-by-side. Dominoes are examples of polyominoes. A pentomino is a set of five squares connected side-by-side. There are twelve pentominoes; however, only eight pentominoes lend themselves to problem solving involving algebra.

Looking at the Algebra Although the pentominoes are all different, the algebra involved is the same for certain groups.

Group 1

a			
b	c	d	e

a			
c	b	d	e

For these two pentominoes, we have a + b = b + c + d + e, or a = c + d + e.

The result shows that the digit representing **b** can be any digit 0 through 9.

The greatest possible sum for a + b is 17, which comes from 8 + 9 or 9 + 8. The smallest possible sum for either pentomino can be obtained with b = 0, a = 6 = 1 + 2 + 3. Therefore c, d, and e are equal to 1, 2, and 3 in any order.

Here are some solutions:

7			
3	0	2	5

8			
1	9	3	4

6			
1	7	2	3

9			
3	1	4	2

Group 2 A second group contains three pentominoes.

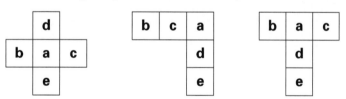

For each, a + b + c = a + d + e, or b + c = d + e.

The greatest sum of 22 occurs when a = 9.

The greatest possible sum for two pairs of digits is 13 = 8 + 5 = 7 + 6.

The least possible sum of 5 occurs when a = 0; therefore, the least possible sum of the two pairs is 5 = 1 + 4 = 2 + 3.

Here are some solutions:

	5	
4	1	7
	6	

3	9	1
		8
		4

7	0	6
		8
		5

DEVELOPING ALGEBRAIC THINKING

PENTOMINO PUZZLES

Teacher's Notes

Group 3 The last group also contains three pentominoes.

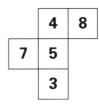

The algebra involves three equations, which are the same for each pentomino.

$a + b = b + c + d = d + e$.

With $a + b = b + c + d$, then $a = c + d$.

Similarly, with $d + e = b + c + d$, then $e = b + c$.

There are two possible arrangements of tiles for the greatest sum.

Since **a** is a single digit, the greatest possible sum for $c + d$ is 9.

This gives $c = 1$ and $d = 8$, with $e = 7$ and $b = 6$, or $c = 2$ and $d = 7$, with $e = 9$ and $b = 6$.

The greatest sum is 15.

The least possible sum for $c + d$ is 3, with $a = 3$, $c = 2$, and $d = 1$.

This arrangement gives $b = 4$ and $e = 6$, with the least possible sum being 7.

Here are some other solutions:

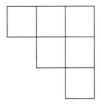 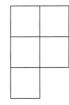 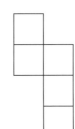

Questions for Discussion Why can't the remaining four pentominoes be used for pentomino puzzles? Observe that in the first three pentominoes shown below, digits would need to be repeated. The fourth pentomino would have only one line of digits.

102 DEVELOPING ALGEBRAIC THINKING © Didax Educational Resources

PENTOMINO PUZZLES

Teacher's Notes

More Ideas The pentominoes in *Group 3* lend themselves to creating puzzles with fixed digits. This allows students to place number tiles in the remaining squares, either to discover the algebraic pattern that exists or to reinforce the algebra developed.

PENTOMINO PUZZLES

Use 5 number tiles so that sum of the digits on each line is the same.

What is the greatest sum you can find? _____

What is the least sum you can find? _____

Record the results in the pentominoes provided.

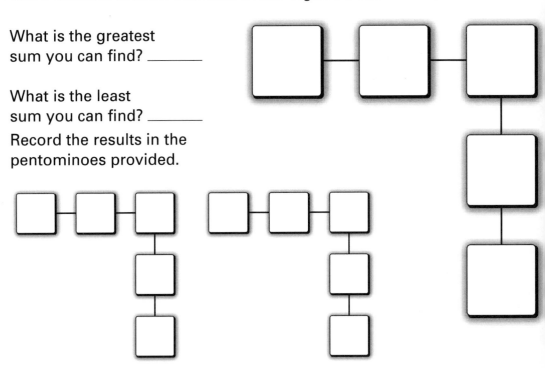

Use 5 number tiles so that sum of the digits on each line is the same.

What is the greatest sum you can find? _____

What is the least sum you can find? _____

Record the results in the pentominoes provided.

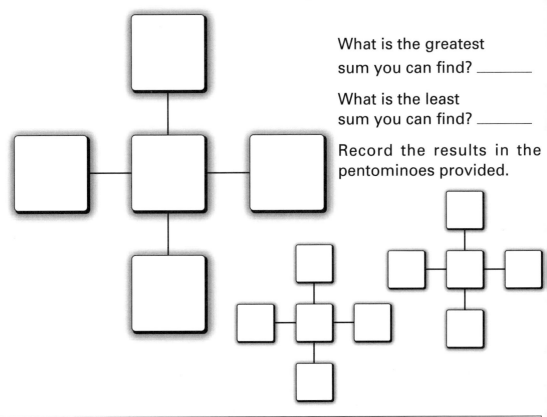

104   DEVELOPING ALGEBRAIC THINKING   © Didax Educational Resources

PENTOMINO PUZZLES

Use 5 number tiles so that sum of the digits on each line is the same.

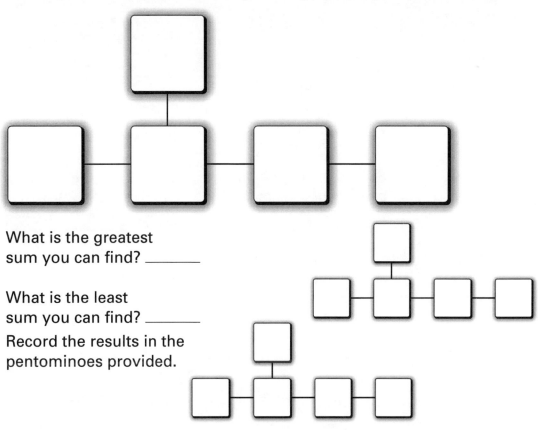

What is the greatest sum you can find? _____

What is the least sum you can find? _____

Record the results in the pentominoes provided.

Use 5 number tiles so that sum of the digits on each line is the same.

What is the greatest sum you can find? _____

What is the least sum you can find? _____

Record the results in the pentominoes provided.

© Didax Educational Resources — DEVELOPING ALGEBRAIC THINKING

PENTOMINO PUZZLES

Use 5 number tiles so that sum of the digits on each line is the same.

What is the greatest sum you can find? _____

What is the least sum you can find? _____

Record the results in the pentominoes provided.

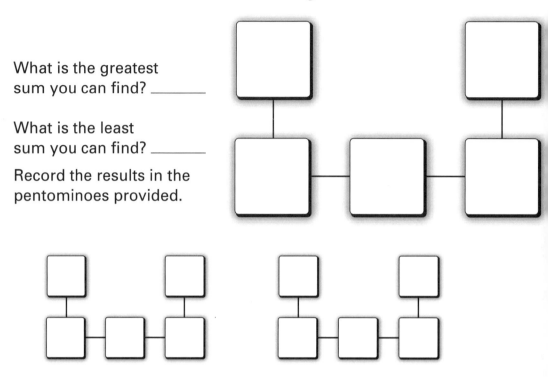

Use 5 number tiles so that sum of the digits on each line is the same.

What is the greatest sum you can find? _____

What is the least sum you can find? _____

Record the results in the pentominoes provided.

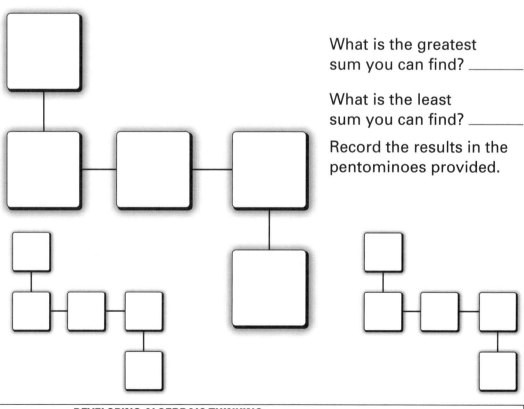

PENTOMINO PUZZLES

Use 5 number tiles so that sum of the digits on each line is the same.

What is the greatest sum you can find? _____

What is the least sum you can find? _____

Record the results in the pentominoes provided.

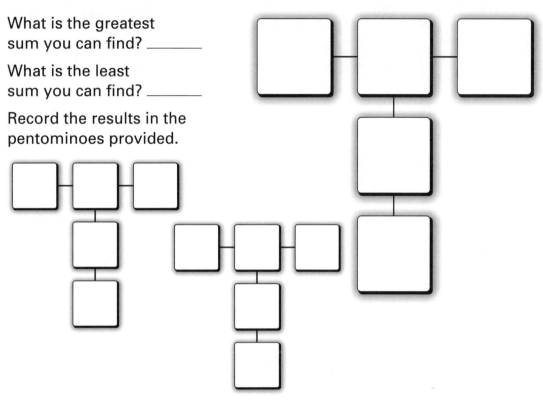

Use 5 number tiles so that sum of the digits on each line is the same.

What is the greatest sum you can find? _____

What is the least sum you can find? _____

Record the results in the pentominoes provided.

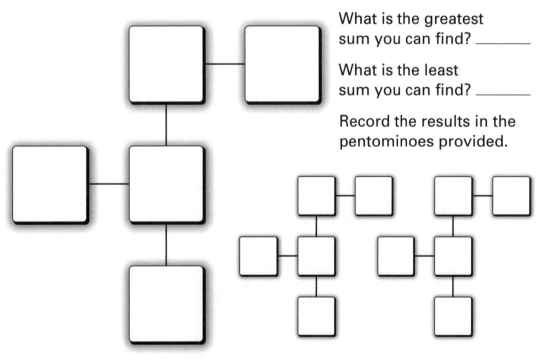

TEACHER'S NOTES

Alphabet Algebra

ALPHABET ALGEBRA

Teacher's Notes

Introduction Every letter of the alphabet can be made into a number tile puzzle, where various lines in the letter all have the same sum. The letters may use some or all of the 10 number tiles. Unlike other sections in this book, where certain digits had to go in certain places, the algebra of *Alphabet Algebra* looks at the relationships that exist among the sums of the digits on the lines.

Looking at the Algebra Again, "Guess and check" is the first-choice strategy for solving *Alphabet Algebra* letters. However, classroom use can focus on variables and equations that develop. In the discussion that follows a letter of the alphabet will have a variable assigned to each digit used in the letter. Equations will be shown and relationships that exist among the variables will be discussed. Possible solutions will be shown.

Alphabet Algebra A

```
      a
   b     e
  c  h  f
  d     g
```

$a + b + c + d = c + h + f = a + e + f + g$
$a + b + d = h + f, \quad c + h = a + e + g$
$b + c + d = e + f + g$

```
      5              7
   7     3        5     3
   6  4  9        4  6  8
   1     2        2     0
```

Alphabet Algebra B

```
  a   f
  b   g
  c h
  d   i
  e   j
```

$a + b + c + d + e = a + f + g + h + c = c + h + i + j + e$
$b + d + e = f + g + h \quad a + f + g = i + j + e$
$a + b + d = h + i + j$

```
  2   7
  4     5
  9 1
  3     8
  6   0
```

Alphabet Algebra C

```
  a h i
  b
  c
  d
  e f g
```

$a + b + c + d + e = a + h + i = e + f + g$
$b + c + d + e = h + i \quad a + b + c + d = f + g$

```
  0 9 5      5 8 4
  8          0
  3          7
  2          3
  1 6 7      2 6 9
```

© Didax Educational Resources   **DEVELOPING ALGEBRAIC THINKING**   111

ALPHABET ALGEBRA

Teacher's Notes

Alphabet Algebra D

a   e
b   f
c   g
d   h

$a + b + c + d = a + e + f + g + h + d$
$b + c = e + f + g + h$

```
5  8        1  6
7     0     7     3
6     2     8     4
1  3        0  2
```

Alphabet Algebra E

a   f g
b
c   h
d
e   i j

$a + b + c + d + e = a + f + g = c + h = e + i + j$
$b + c + d + e = f + g \quad a + b + c + d = i + j$
$a + b + d + e = h$

```
1  9  3        0  5  8
4              4
6     7        6     7
2              2
0  8  5        1  3  9
```

Alphabet Algebra F

a   f g h
b
c   i j
d
e

$a + b + c + d + e = a + f + g + h = c + i + j$
$b + c + d + e = f + g + h \quad a + b + d + e = i + j$

```
4  3  1  9     0  6  8  2
5              5
2     8  7     3     9  4
6              7
0              1
```

Alphabet Algebra G

a   i j
b
c
d   h
e   f g

$a + b + c + d + e = a + i + j = e + f + g = h + g$
$b + c + d + e = i + j \quad e + f = h$
$a + b + c + d = f + g$

```
0  6  8        2  9  3
7              7
4              4
1     5        0     6
2  3  9        1  5  8
```

Alphabet Algebra H

a   d
b   g   e
c   f

$a + b + c = b + g + e = d + e + f$
$a + c = g + e \quad\quad b + g = d + f$

```
6     5        5     2
3  9  1        4  7  1
4     7        3     9
```

DEVELOPING ALGEBRAIC THINKING

ALPHABET ALGEBRA

Teacher's Notes

Alphabet Algebra I

```
f a g
  b
  c
  d
h e i
```

a + b + c + d + e = f + a + g = h + e + i
b + c + d + e = f + g   a + b + c + d = h + i

```
9 2 4          9 0 5
  7              2
  5              3
  0              8
6 1 8          7 1 6
```

Alphabet Algebra J

```
    a
    b
    c
h   d
g f e
```

a + b + c + d + e = e + f + g = g + h
a + b + c + d = f + g   e + f = h

```
    1              7
    3              4
    6              0
  9 2            6 2
  8 4 5          8 5 1
```

Alphabet Algebra K

```
a   g
b   f
c
d   h
e     i
```

a + b + c + d + e = c + f + g = c + h + i
a + b + d + e = f + g = h + i

Let f + g be even. Divide that sum by 2. Find two pairs with that new sum. These four digits are on the vertical line.

```
7   5          3   1
0   9          2   9
2              7
3   8          0   4
4     6        5     6
```

Alphabet Algebra L

```
a
b
c
d
e f g h
```

a + b + c + d + e = e + f + g + h
a + b + c + d = f + g + h

Each time this equation is solved, three different solutions are created.

```
0
3
5
7
9 8 6 1
```

Replacing the 9-tile with 2 or 4 gives a different solution.

DEVELOPING ALGEBRAIC THINKING

ALPHABET ALGEBRA

Teacher's Notes

Alphabet Algebra M

```
a       g
b  d f  h
c    e    i
```

$a + b + c = a + d + e = e + f + g = g + h + i$
$b + c = d + e$ $a + d = f + g$ $e + f = h + i$

0   1   3   6
7 5 4 3   8 5 2 9
6 8 9   4 7 0

Alphabet Algebra N

```
a       f
b   e   g
c       h
d       i
```

$a + b + c + d = a + e + i = f + g + h + i$
$b + c + d = e + i$ $a + e = f + g + h$

5   4   8   9
3 8 9   3 7 5
6   0   6   1
1   2   2   4

Alphabet Algebra O

```
a  b  c
h     d
g  f  e
```

$a + b + c = c + d + e = e + f + g = g + h + a$
$a + b = d + e$ $c + d = f + g$ $e + f = h + a$

7 0 5   8 1 6
2   6   3   7
3 8 1   4 9 2

Alphabet Algebra P

```
a  f  g
b     h
c  j  i
d
e
```

$a + b + c + d + e = a + f + g = g + h + i = i + j + c$
$b + c + d + e = f + g$ $a + f = h + i$ $g + h = j + c$

9 6 2   7 6 2
3   7   4   8
4 5 8   1 9 5
1   0
0   3

Alphabet Algebra Q

```
a  h  g
b     f
c  d  e
         i
```

$a + b + c = c + d + e = e + f + g = g + h + a$
$= e + i$
$a + b = d + e$ $c + d = f + g$ $e + f = h + a$
$c + d = i$ $f + g = i$

8 5 3   7 5 1
2   4   0   8
6 1 9   6 3 4
      7   9
```
114    DEVELOPING ALGEBRAIC THINKING    © Didax Educational Resources
```

ALPHABET ALGEBRA

Teacher's Notes

Alphabet Algebra R

```
a    f
b      g
c    h
d        i
e        j
```

$a + b + c + d + e = a + f + g + h + c = h + i + j$
$b + d + e = f + g + h \quad a + f + g + c = i + j$

```
1   5        2   0
0     2      9     7
4   7        4   6
6     9      3     8
8     3      1     5
```

Alphabet Algebra S

```
b   a
c
d   e
    f
h   g
```

$a + b = b + c + d = d + e = e + f + g = g + h$
$a = c + d \quad\quad b + c = e \quad\quad d = f + g \quad\quad e + f = h$

```
2   9        5   8
5            1
4   7        7   6
    1            3
8   3        9   4
```

Alphabet Algebra T

```
a b c d e
    f
    g
    h
```

$a + b + c + d + e = c + f + g + h$
$a + b + d + e = f + g + h$
Each time this equation is solved, three different solutions are created. There are two tiles remaining which can be interchanged for c.

```
8 2 9 4 0       8 1 9 4 2
    7               3
    6               7
    1               5
```

The 9 could be replaced by 3 or 5.
(The 9 could be replaced by 0 or 6.)

Alphabet Algebra U

```
a       i
b       h
c       g
d   e   f
```

$a + b + c + d = d + e + f = f + g + h + i$
$a + b + c = e + f \quad\quad d + e = g + h + i$
The sum of the tiles range from 7 to 16.

```
7       6       4       5       7       8
4       5       2       8       3       2
1       0       1       3       0       1
8       3  9    9  7  0         5  6  4
```

© Didax Educational Resources — DEVELOPING ALGEBRAIC THINKING

ALPHABET ALGEBRA

Teacher's Notes

Alphabet Algebra V

```
    a     g
    b   f
    c  e
      d
```

$a + b + c + d = d + e + f + g$
$a + b + c = e + f + g$

Each time this equation is solved, four different solutions are created. There are three tiles remaining which can be interchanged for d.

```
  6     3
   0   4
    2 1
     7
```

The 7 could be replaced with 5, 8 or 9.

Alphabet Algebra W

```
  a       i
    e
  b   d f   h
  c       g
```

$a + b + c = c + d + e = e + f + g = g + h + i$
$a + b = d + e \qquad c + d = f + g \qquad e + f = h + i$

```
  6         9         4         0
      8              7
  7     3         8     9
    5 4             5 2
  0     1         3     6
```

Alphabet Algebra X

```
  a       i
   b    h
     c
    g d
  f     e
```

$a + b + c + d + e = f + g + c + h + i$
$(a + b) + (d + e) = (f + g) + (h + i)$

Each time this equation is solved, two different solutions are created.

```
     8   4
    2   6
       0
    3   1
   7     9
```

Alphabet Algebra Y

```
  d     g
   e  f
     a
     b
     c
```

$a + b + c = d + e + a = a + f + g$
$b + c = d + e = f + g$

Each time this set of equations is solved, four different solutions are created.

```
   2   0
  3   5
    6
    1
    4
```

The 6-tile could be replaced by 7, 8 or 9.

Alphabet Algebra Z

```
  a b c d
      e
      f
  g h i j
```

$a + b + c + d = d + e + f + g = g + h + i + j$
$a + b + c = e + f + g \qquad d + e + f = h + i + j$

```
  2 9 3 5        9 5 1 2
      6              3
      1              8
  7 0 8 4        4 7 0 6
```

DEVELOPING ALGEBRAIC THINKING

ALPHABET ALGEBRA A

Use any 8 number tiles 0 through 9 in the squares so that the sum of each line is the same.

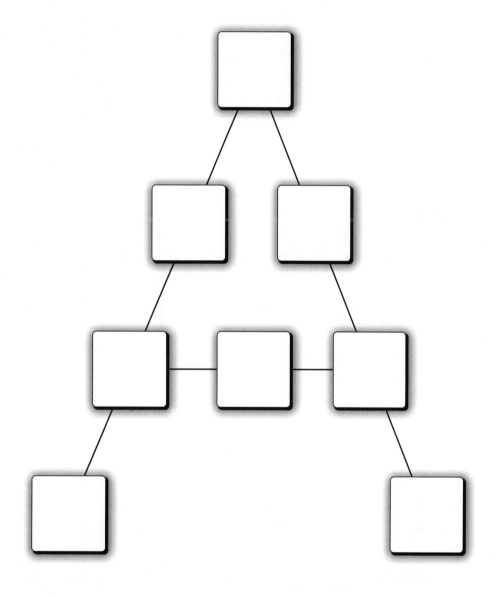

ALPHABET ALGEBRA B

Use all 10 number tiles 0 through 9 in the squares so that the sum of each line is the same.

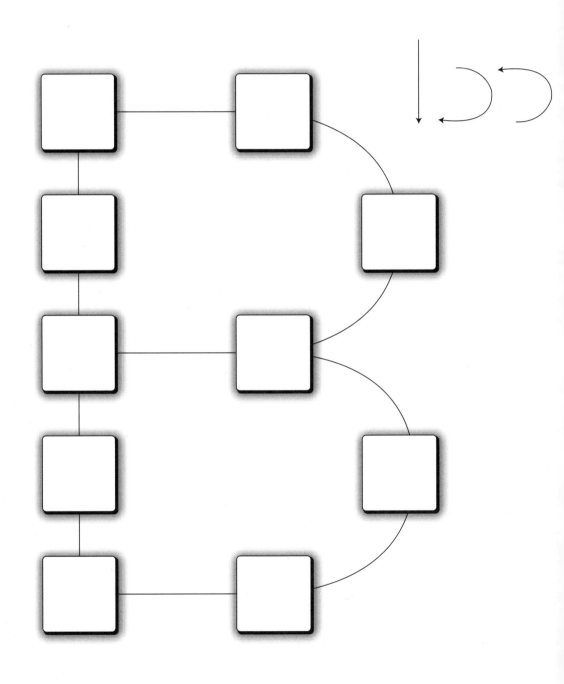

ALPHABET ALGEBRA C

Use any 9 number tiles 0 through 9 in the squares so that the sum of each line is the same.

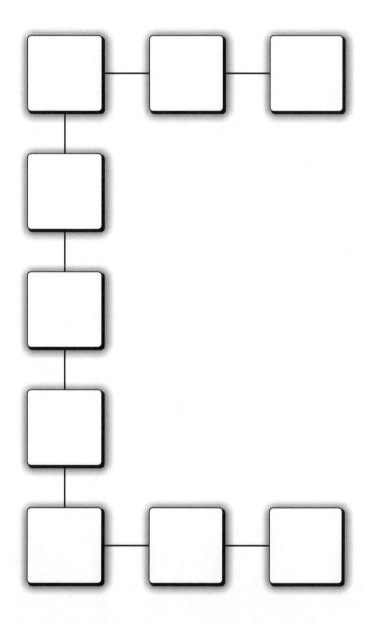

ALPHABET ALGEBRA D

Use any 8 number tiles 0 through 9 in the squares so that the sum of each line is the same.

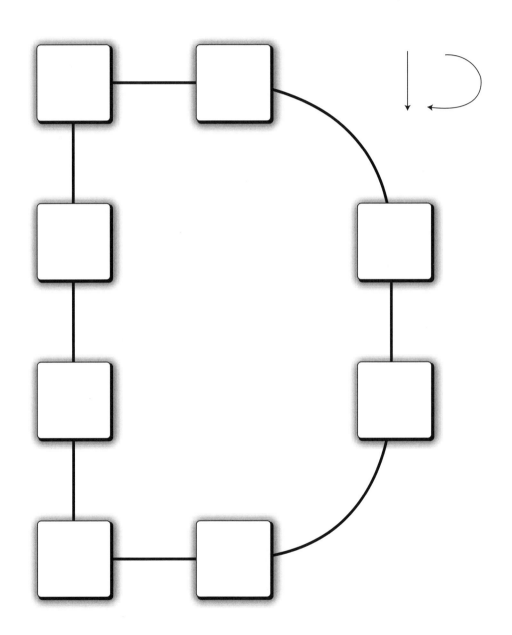

ALPHABET ALGEBRA E

Use all 10 number tiles 0 through 9 in the squares so that the sum of each line is the same.

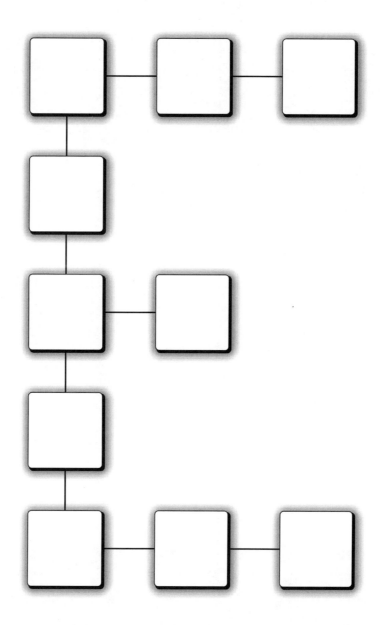

ALPHABET ALGEBRA F

Use all 10 number tiles 0 through 9 in the squares so that the sum of each line is the same.

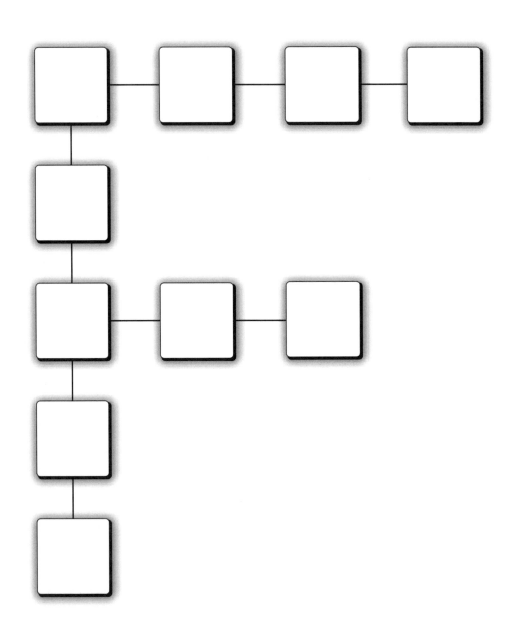

ALPHABET ALGEBRA G

Use all 10 number tiles 0 through 9 in the squares so that the sum of each line is the same.

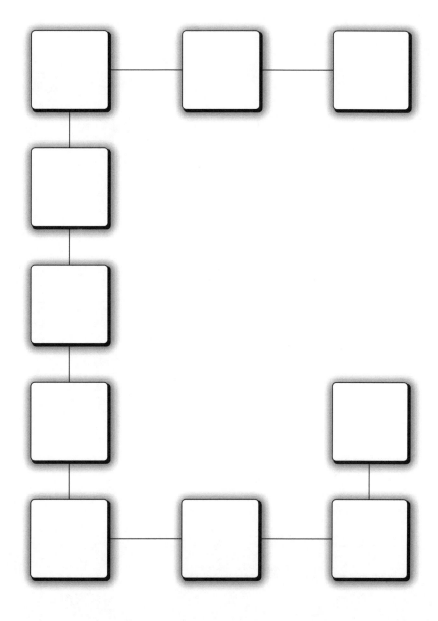

ALPHABET ALGEBRA H

Use any 7 number tiles 0 through 9 in the squares so that the sum of each line is the same.

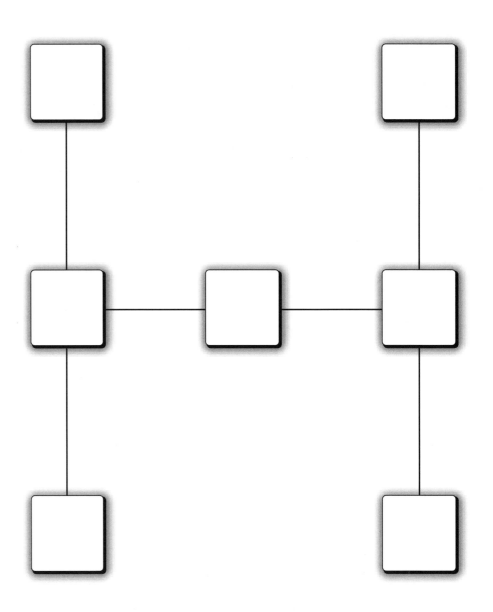

ALPHABET ALGEBRA I

Use any 9 number tiles 0 through 9 in the squares so that the sum of each line is the same.

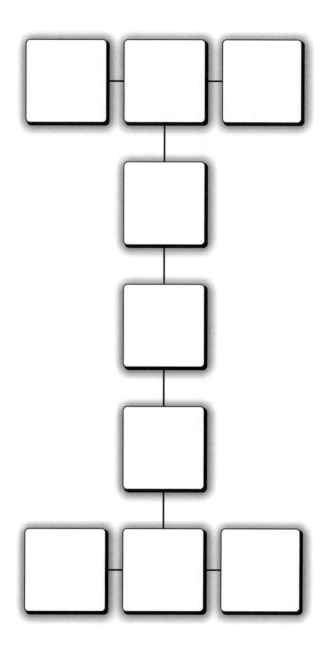

DEVELOPING ALGEBRAIC THINKING

ALPHABET ALGEBRA J

Use any 8 number tiles 0 through 9 in the squares so that the sum of each line is the same.

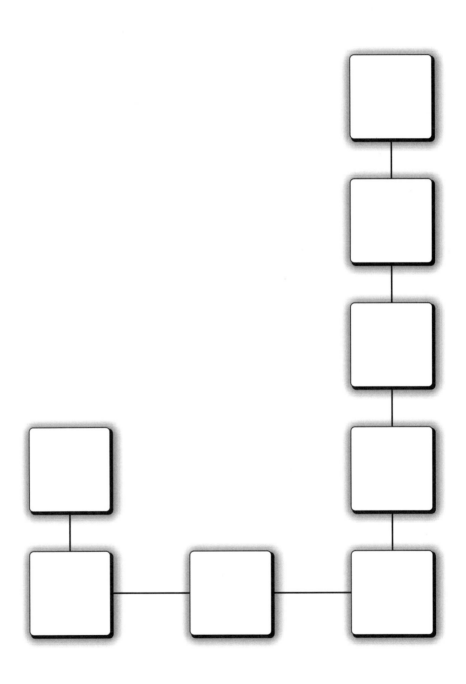

ALPHABET ALGEBRA K

Use any 9 number tiles 0 through 9 in the squares so that the sum of each line is the same.

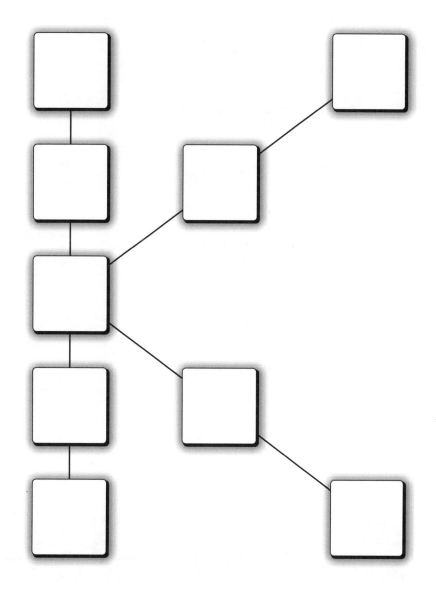

DEVELOPING ALGEBRAIC THINKING

ALPHABET ALGEBRA L

Use any 8 number tiles 0 through 9 in the squares so that the sum of each line is the same.

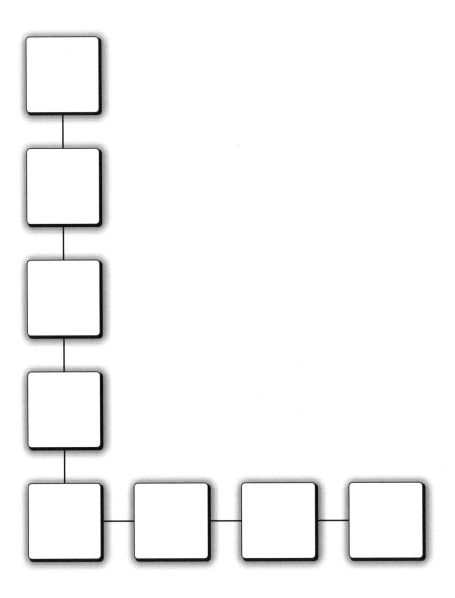

ALPHABET ALGEBRA M

Use any 9 number tiles 0 through 9 in the squares so that the sum of each line is the same.

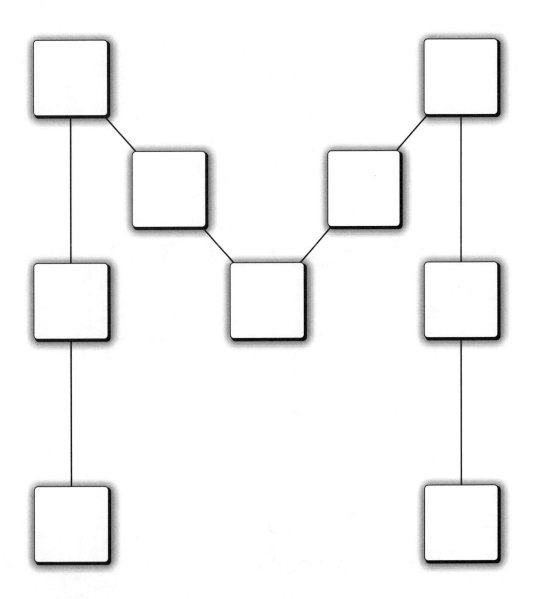

© Didax Educational Resources — DEVELOPING ALGEBRAIC THINKING

ALPHABET ALGEBRA N

Use any 9 number tiles 0 through 9 in the squares so that the sum of each line is the same.

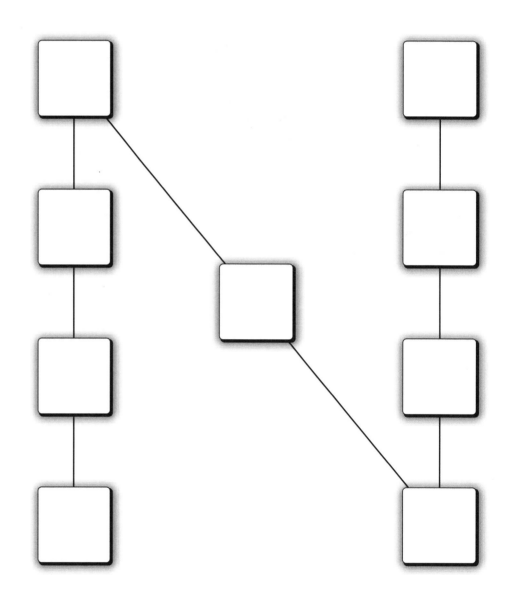

ALPHABET ALGEBRA O

Use any 8 number tiles 0 through 9 in the squares so that the sum of each line is the same.

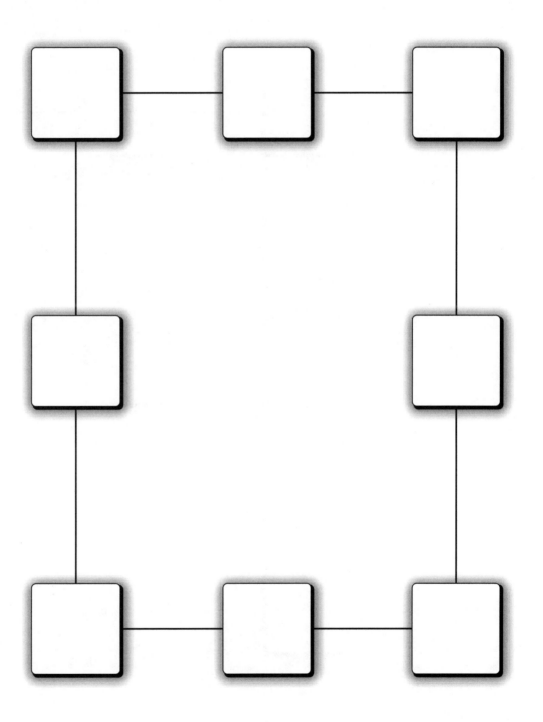

ALPHABET ALGEBRA P

Use all 10 number tiles 0 through 9 in the squares so that the sum of each line is the same.

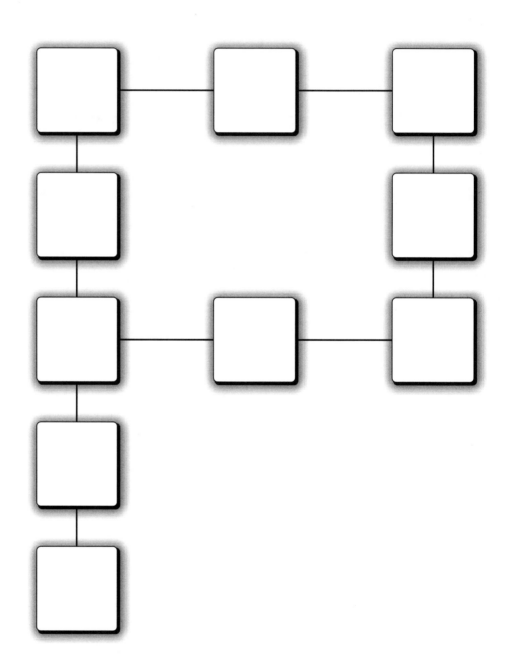

ALPHABET ALGEBRA Q

Use any 9 number tiles 0 through 9 in the squares so that the sum of each line is the same.

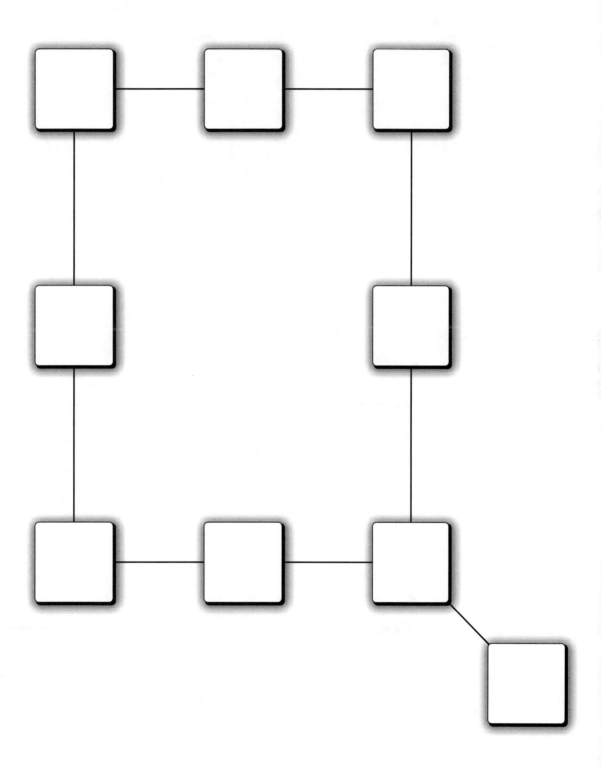

ALPHABET ALGEBRA R

Use all 10 number tiles 0 through 9 in the squares so that the sum of each line is the same.

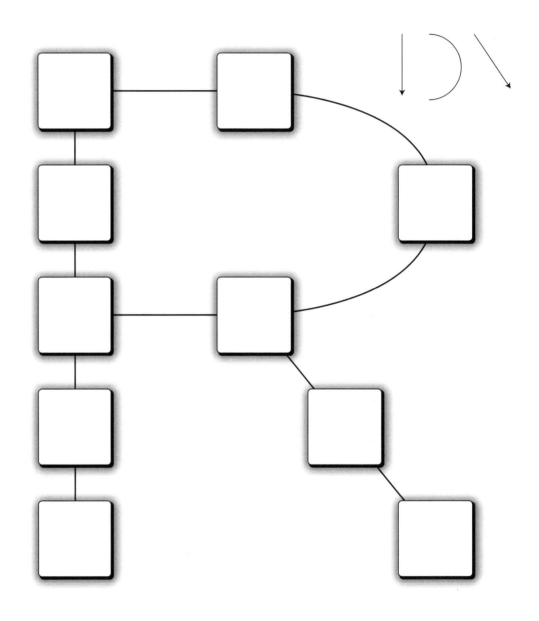

134 DEVELOPING ALGEBRAIC THINKING © Didax Educational Resources

ALPHABET ALGEBRA S

Use any 8 number tiles 0 through 9 in the squares so that the sum of each line is the same.

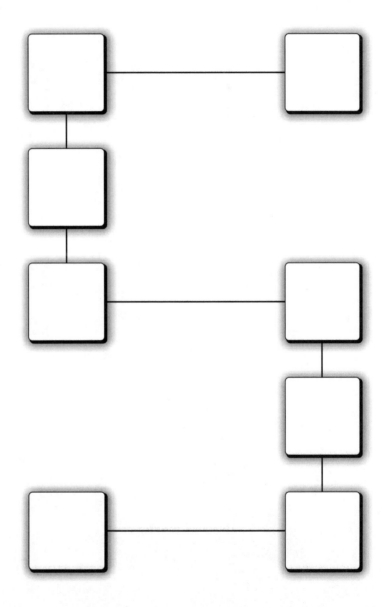

ALPHABET ALGEBRA T

Use any 8 number tiles 0 through 9 in the squares so that the sum of each line is the same.

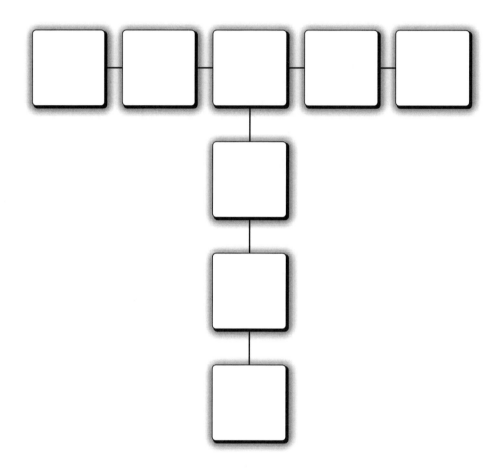

ALPHABET ALGEBRA U

Use any 9 number tiles 0 through 9 in the squares so that the sum of each line is the same.

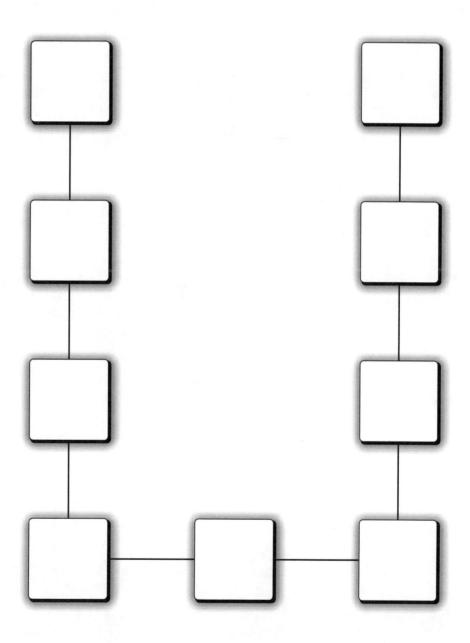

ALPHABET ALGEBRA V

Use any 7 number tiles 0 through 9 in the squares so that the sum of each line is the same.

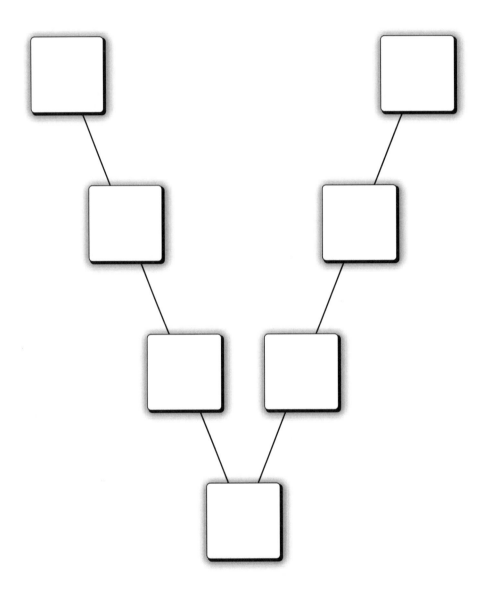

ALPHABET ALGEBRA W

Use any 9 number tiles 0 through 9 in the squares so that the sum of each line is the same.

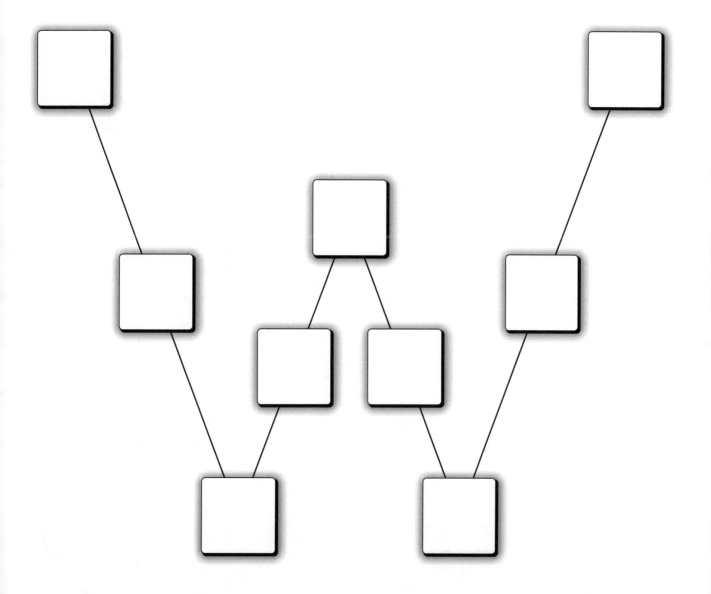

ALPHABET ALGEBRA X

Use any 9 number tiles 0 through 9 in the squares so that the sum of each line is the same.

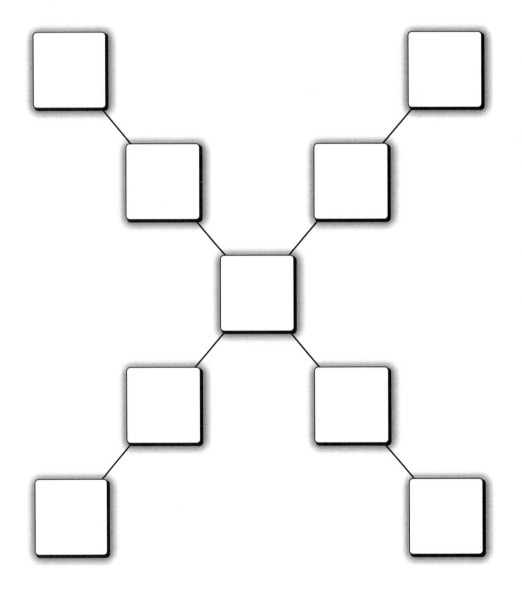

ALPHABET ALGEBRA Y

Use any 7 number tiles 0 through 9 in the squares so that the sum of each line is the same.

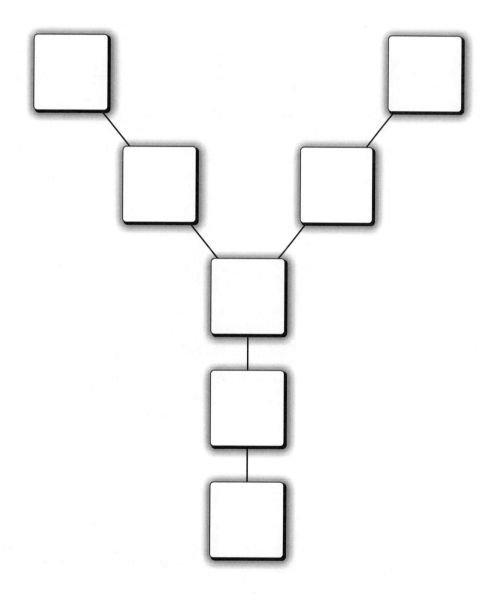

ALPHABET ALGEBRA Z

Use all 10 number tiles 0 through 9 in the squares so that the sum of each line is the same.

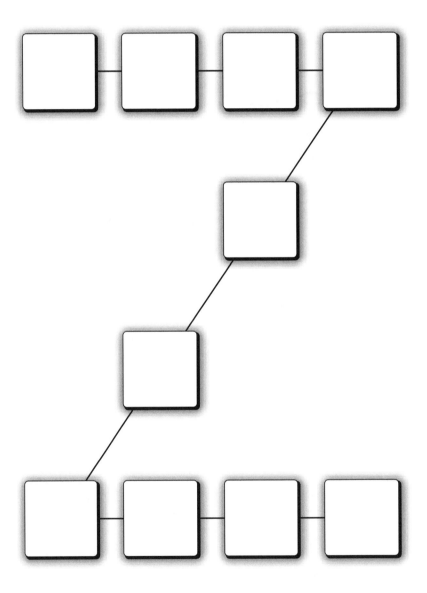

NUMBER TILES

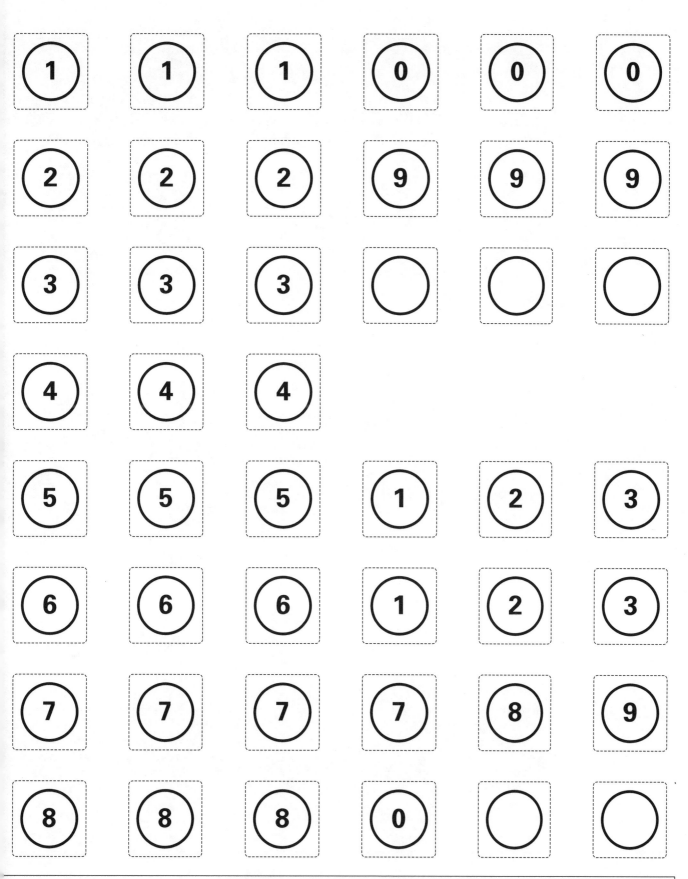

DEVELOPING ALGEBRAIC THINKING

TEACHER'S NOTES